Technical Writing Essentials

Technical Writing Essentials

Michael H. Markel
Drexel University

ST. MARTIN'S PRESS
New York

Library of Congress Catalog Card Number: 87-060508

Manufactured in the United States of America.
21098
fedcba

For information, write:
St. Martin's Press, Inc.
175 Fifth Avenue
New York, NY 10010

Interior design: Leon Bolognese
Graphics: G&H Soho
Cover: Joel Weltman

ISBN: 0-312-00736-1
Instructor's Edition ISBN: 0-312-01287-X

Acknowledgments
p. 64: Flowchart from *Solar Energy Utilization* by Tim Michels. Copyright © 1979 by Van Nostrand Reinhold. All rights reserved.

p. 65: Flowchart reprinted with the permission of the Institute of Electrical and Electronics Engineers, Inc. © 1985 IEEE. By Neng F. Yao, "The Computer-Aided Programming System—A Friendly Programming Environment," in *IEEE Micro*, vol. 5, no. 2 (April 1985), 12.

pp. 66–67: Diagrams from *Complete Do-It-Yourself Manual* © 1973 by The Reader's Digest Association, Inc. Reprinted by permission of the publisher.

TO DAVID

Preface

The purpose of this book is to help you improve the effectiveness of your writing on the job. I want to make it easier for you to write and easier for your readers to understand what you have written.

Everyone has experienced the trauma of staring at a blank sheet of paper and being unable to fill it with coherent writing. But when you're getting *paid* to write something, and your boss wanted it yesterday, the word *deadline* takes on a new and personal significance.

The common advice is to start to write. Don't fight the words—simply let them flow. Somewhere between your brain and the pen, all your problems will be solved.

This advice sounds appealing, but in the real world few people can let the words flow without having them sound as if they were spilled. Writing is a rational process, not a mysterious and magical art. Good writing requires a logical approach that is a lot closer to plain common sense than to inspiration. This book describes a simple but effective approach for any kind of writing task on the job.

In my experience working with thousands of professionals in every field of engineering, science, and business, I have found a remarkably similar set of writing problems:

1. The writer finds it hard to get started on the assignment and wastes a good deal of time.
2. The document ends up too long; it includes all the technical details and doesn't get directly to the important points. The reader is frustrated.
3. The document does not clearly reflect what the writer wanted to say; minor points receive too much emphasis, and major points slip by almost unnoticed.

The best way to prevent these problems is, first, to figure out who is going to read what you've written, and why you are writing. Next, you brainstorm and do the necessary research. With these preliminary steps out of the way, you are ready to outline and draft the document, whether it is a letter, memo, manual, or report. Finally, you revise the draft to make sure it communicates the message you intend.

The emphasis of this book is on creating a structure that reflects your purpose—that is, what you're trying to accomplish in the writing. In most cases, this means devising a format that is not only clear, comprehensive, and concise, but also accessible, so that the various readers can easily find what they're looking for. Accessibility is a concept that covers every aspect of writing, from preliminary summaries for readers who just want the "bottom line," to paragraphs that announce their main point in the first sentence, to sentences that focus directly on the important information.

In the discussion of style—covering such topics as grammar, usage, and paragraphing—I stick to the essentials, the rules that make a difference. I don't have time for the small points, because in the real world, writers don't. A perfect memo that requires three hours of polishing would not be cost effective.

After a discussion of the common graphic aids used in presenting technical information, I cover two of the common formats: letters and memos. The discussion of reports begins with a chapter describing the basic elements of reports. Then I treat proposals, progress reports, and completion reports. The next chapter describes oral presentations of technical information. The final chapter describes the job-application process. The appendixes include a handbook of style, punctuation, and mechanics; a discussion of how to document sources; a discussion of how word processing can help you in every stage of the writing process; and a selected bibliography.

No book, no matter how good it is, can turn you into an excellent writer. You have to write and keep writing. The approach described in this book, however, has given many students and professionals the skills and self-confidence to reduce the time and aggravation that go into writing and to improve dramatically the effectiveness of the finished product. I think it will work for you, too.

I am indebted to the staff at St. Martin's Press, especially acquiring editor Susan Anker and senior project editor Patricia Mansfield, and to Karen Mitchell. The book has been improved substantially by the suggestions of the reviewers: Deborah J. Barrett, Houston Baptist University; Susan Tyler Hitchcock, University of Virginia; David Farkas, University of Washington; and Janet Evans Worthington, West Virginia Institute of Technology.

My greatest debt, as always, is to my wife, Rita.

Michael H. Markel

Contents

1

Introduction to Technical Writing

The Role of Technical Writing in Business and Industry

Characteristics of Effective Technical Writing

Clarity
Accuracy
Comprehensiveness
Accessibility
Conciseness
Correctness

Technical writing is nonfiction writing about a technical subject, addressed to a particular audience to achieve a particular purpose.

Technical writing is generally thought of as concerning subjects such as the sciences, technology, and business. However, a broad definition of technical-writing subjects can include the arts. The history of sculpture, for instance, can be written about as a technical subject, as can a particular poet's use of the Italian sonnet form.

What is "nonfiction writing"? It is any writing that is factual or that relies on facts. Except for any typographical errors, a telephone book is factual: the names, addresses, and phone numbers are accurate. A report on techniques for getting oil out of undersea wells is also factual if it describes methods that are being used or that have been used. Speculations on techniques that will be used in the future are considered nonfiction, too, as long as the writers indicate they are speculating and base their predictions on the best available evidence and on clear thinking.

What is a "particular audience"? An audience could be one person, several people, or even thousands. The actual number does not matter. What is important is that the writer is intent on reaching *that* audience. For this reason, the writer thinks carefully about that audience before starting to plan a document. What the readers already know about the subject, the way they feel about it, how they are going to read and use the information these and many more questions must be answered before the writer can determine how to plan the document.

What is a "particular purpose"? Every document has its own purpose or purposes, of course, but in general there are two kinds: to explain something to someone and to affect that person's attitude about something. For instance, a set of instructions on how to install an electric garage-door opener has an obvious purpose: to explain to the reader how to install the device safely and easily. A status report written by the manager of the reference department at a large library has as its purpose explaining what happened in the reference department during the period the report covers: what special and routine tasks were accomplished, or what events occurred.

In addition to explaining something, most technical documents are meant to affect the reader's attitudes. For instance, a corporate annual report communicates what happened in the corporation that year. But it has other purposes: to suggest that the corporation is well managed and healthy and to inspire confidence among the stockholders.

On a more personal level, every writer will benefit by leaving the reader with the impression that the work being reported was performed compe-

tently and that the document itself meets the highest standards of effective communication.

The Role of Technical Writing in Business and Industry

The working world is so complex that virtually everything that occurs has to be documented. The need to write effectively is becoming more and more important.

Your first technical-writing task will arise before you start your first professional job: writing a persuasive job-application letter and résumé. The potential employer will want to know as much as possible about you: where you went to school, what you studied, how well you did, what jobs you held, what extracurricular activities you participated in, and so forth. This information, obviously, is what you are trying to communicate. But your letter and résumé fulfill another important function, too. They tell your reader whether you take pride in your work. A poorly organized résumé or an inadequately developed letter says either that you are not first rate or that you're not trying. Either way, your chances of getting an interview are poor. However, if your application materials are carefully thought out and presented, the potential employer will infer that all your work will be impressive. You'll make it to the next round.

On the job, you will write every day. You will write letters to clients and suppliers, and memos to subordinates and supervisors. If you go on a business trip, you probably will write a trip report when you return; if the report is clear and informative, your supervisors will be encouraged to give you more responsibility (and status and money).

The more important you become at an organization, the more you will write. If you need some equipment or other resources, you will write a proposal describing what you want and why. If you make a persuasive case, your request is likely to be granted. As you work on the project, you will write progress reports describing where you stand, the problems you've encountered and how you've handled them, and how the original goals of the project will be affected. When the project is complete, you will submit a completion report describing the problem, methods, results, conclusions, and recommendations.

The world of work is the world of technical writing; writing plays an important role in every phase of routine operations and special projects.

Characteristics of Effective Technical Writing

Technical writing is meant to get a job done. Everything else is secondary. If the writing style is interesting, so much the better. But keep in mind the six basic characteristics of effective technical writing:

1. clarity
2. accuracy
3. comprehensiveness
4. accessibility
5. conciseness
6. correctness

Clarity

The most important characteristic of technical writing is clarity. The written document must convey a single meaning that the reader can understand easily.

Unclear technical writing is expensive. A typical letter costs over ten dollars in labor and materials; the average page of a report, over fifty dollars. But these dollar figures are misleadingly low, because of the cooperative nature of most projects today. While an unclear document is being rewritten, a whole team of people can be waiting. This, of course, is expensive. Or, even worse, a team can start to work on the basis of the information contained in an unclear document. Wrong quantities of materials are purchased, construction begins in the wrong location, and so forth.

Unclear technical writing can also be dangerous. Poorly written warnings on medication bottles are a common example, as are unclear instructions on how to operate machinery. A carelessly drafted building code tempts contractors to save money by using inferior materials or methods.

Accuracy

All the problems that can result from unclear writing can also be caused by inaccurate writing.

Accuracy is a simple concept in one sense: you must record your facts carefully. If you mean to write "2,000," don't write "20,000." If you mean to refer to "Figure 3-1," don't refer to "Figure 3-2." Inaccuracies will at least annoy and confuse your reader. They can also be dangerous.

In another sense, accuracy is a more difficult concept. Technical writing must be as objective and unbiased as a scientific experiment. If your readers

suspect you are slanting the information—by overstating or omitting a particular point—they will doubt the validity of the whole document. Technical writing must be effective by virtue of its clarity and organization, but it must also be reasonable, fair, and honest.

Comprehensiveness

A comprehensive technical document provides all the information the readers will need. It provides the background so that readers who are unfamiliar with the project will be able to understand the problem or opportunity that led to the project. It includes a clear description of the methods the writer used to carry out the project, as well as a complete statement of the principal findings—the results and any conclusions and recommendations.

Comprehensiveness is crucial for two reasons. First, the people who will act on the document need a complete, self-contained discussion so that they can apply the information efficiently and safely. And second, the document will function as the official company record of the project, from its inception to its completion.

For example, a scientific article reporting on an experiment comparing the reaction of a new strain of bacterium to two different compounds will not be considered for publication unless the writer has described fully the methods used in the experiment. Because other scientists want to be able to replicate the researcher's methods, every detail, including the names of the companies from which the researcher obtained all the materials, is included.

Or consider a report recommending that a company network its computers. The company will probably want to have the recommendations analyzed in detail before committing itself to such an expensive and important project. The team charged with studying the report needs all the details. If the recommendations are implemented, the company needs a single complete source of information so that if changes have to be made several months or years later, there will be a complete description of what was done and why.

Accessibility

An accessible document is one that is structured so that the readers can easily locate the information they seek. Most technical documents are made up of small, independent sections. Some readers are interested in only one or in several of the sections. Other readers might read most of them. But relatively few people will read the whole document from start to finish, like a novel.

Therefore, you should make your document easy to access by creating self-contained discussions and by using headings and lists (see Chapter 3) and, for reports, a detailed table of contents (see Chapter 6).

Conciseness

Technical writing is meant to be useful. For a document to be useful, people have to read it. A short document is much more likely to be read than a long document. Therefore, your writing should be as concise as you can make it without sacrificing the other criteria of effective writing.

One simple way to shorten a document is to get rid of the long words and phrases. Instead of writing, "The lower inflation rate must be taken into consideration," just say, "The lower inflation must be considered." Before writing, "The fact of the matter is that . . .," ask yourself if the phrase says anything at all.

The real enemy of conciseness, however, isn't the individual word or phrase. Rather, it is the wrong assumption that long documents are better than short ones. We pick up this bad idea in school, where our assignments are required to be "at least" 1,000 words or 10 pages or some other length. Rarely, if ever, do we hear that an assignment must be no longer than three pages. So, we think that our readers like long documents. They don't. Just like you, they like short documents.

Correctness

Good technical writing is correct: it observes the conventions of spelling, grammar, punctuation, and usage.

Some of the conventions are important in an obvious way: if you write, "While feeding on the algae, the researchers captured the fish," you've got the researchers eating algae. Most of the conventions, however, are important because they make you look professional. If your document is full of careless writing errors, your readers will begin to doubt the accuracy of your technical information. Although some very bright people can't spell, most of them use a dictionary or a spelling-checker program.

Technical writing is meant to fulfill a mission: to convey information to a particular audience or to affect that audience's attitudes in a particular way. To accomplish these goals, a document must be clear, accurate, complete, and easy to access. It must be economical and correct. The writer must be invisible. The only evidence of his or her hard work is a document that works—without the writer's being there to explain it.

2

The Writing Process

The highly complex mental functions involved in writing are not completely understood. One thing, however, is certain: just as no two persons think alike, no two persons use the same process in writing. And the same person might well use different techniques on different occasions. In one sense, therefore, the term *writing process* is misleading. There are really an infinite number of writing processes.

Yet, research conducted over the last decade has shown that, despite the considerable variations, most persons do their best work when they treat writing as a process consisting of smaller tasks. This chapter treats writing as a five-part process:

1. preparing
2. doing research
3. organizing
4. drafting
5. revising

Many technical persons dislike writing and try to finish it as quickly as possible. They treat writing as a two-part process: writing and typing. One morning, they start with a clean sheet of paper, write "I. Introduction" at the top, and hope the rest will follow. No wonder they dislike writing. A few geniuses can write like that, but the rest of us find that it doesn't work.

We end up staring at the blank sheet for half an hour or so, unable to think of what to say. After another half-hour, we have created a miserable little paragraph. We then spend the rest of the morning trying to turn that paragraph into coherent English. We change a little here and there, clean up the grammar and punctuation, and consult the thesaurus. Finally, it's time to go to lunch. We return, happy that we have at least created a good paragraph. Then we reread what we've written. We realize that all we have done is waste a morning. We start to panic, which makes it all the more difficult to continue writing.

Some writers use a somewhat more effective process. They realize they need some sort of outline to follow. But they do only a halfhearted job on the outline, scribbling a few words on a pad. Outlining has bad associations for most people. All of us at one time or another have had to hand in outlines to teachers who seemed more interested in correcting the format—the Roman numerals and the indention—than in suggesting ways to help write the papers. And sometimes the outline assignment was a punishment for having written a poorly organized paper. "Take this home and outline it. Then you'll see what a mess it is!"

The writing process described in this chapter makes writing less mysterious and more like the other kinds of technical tasks that you carry out all the time.

Preparing

Chapter 1 defines technical writing as nonfiction about a technical subject, addressed to a particular audience to achieve a particular purpose. In other words, the content and form of any technical document are determined by the situation that calls for that document. This writing situation is made up of the person or persons the document is intended for and the writer's reasons for creating the document.

Audience and purpose are not concerns unique to technical writing. Most communication situations in the working world are determined by the same two factors. For instance, when you write a job-application letter and résumé, your audience is the personnel officer or someone else responsible for hiring you. Your purpose is to impress that person favorably so that you are invited to a job interview.

Once you have determined your audience and purpose, you must analyze each one to decide the content and form of your document. The best way to begin is by analyzing your audience, because you must understand your audience before you can understand what you want them to know or think after they have finished reading your document.

Analyzing Your Audience

Identifying and analyzing your audience can be difficult. You have to put yourself in the position of people you might not know reading something you haven't yet written. Most writers want to concentrate first on content—the nuts and bolts of what they have to say. They want to write something first and shape it later.

Resisting this temptation is crucial. Having written primarily for teachers, students do not automatically think much about their audience. In most cases, students have some idea of what their teachers want to read. In addition, teachers establish guidelines and expectations for their assignments, and they usually understand what their students are trying to say. The typical teacher is a known quantity. In business and industry, however, you will often have to write to different audiences, each of which has a different level of knowledge of your area of expertise. And these different audiences might well have very different purposes in reading what you have written.

How do you analyze your audience? Sometimes it's difficult because you don't know who they will be. You might be writing a report for your supervisor, and he or she is likely to distribute it—but you don't know to whom. Or you might be addressing an audience of several hundred. Even though these many readers share some characteristics, they cannot be thought of as a unified group.

Despite these difficulties, however, often it is possible to analyze an audience accurately. First, try to determine whether one person or several will be reading. For every reader whom you can identify, ask yourself a set of questions such as the following:

1. What is the person's name and job title?
2. What are the person's chief responsibilities on the job?
3. What is the person's educational background?
4. What does the person already know about the subject?
5. What will the person do with the document? file it? skim it? read only a portion of it? study it carefully? modify it and submit it to another reader? work from it in attempting to implement its recommendations?
6. What do you know about the person's likes and dislikes that might affect his or her reaction to the document?

Writing out profiles of your audience is a good idea because it forces you to be as specific as possible. Following is an example.

You are writing a monthly status report on the activities of your mechanical engineering group. Your reader is Bill Harpole, a 42-year-old manager of engineering operations. In this position, he supervises the operations of four different groups. Bill is trained as a civil engineer, with a graduate degree in business administration. He is reasonably knowledgeable in your area: for six years he worked as a project manager and interacted with civil engineers daily.

Bill's task is to take the information from the four status reports, summarize it, and submit it to his supervisor. You know little about Bill's supervisor, Karen Stahl, director of operations, except that she is trained in operations research, the study of systems. Bill has told you not to worry about her as you write to him.

Because Bill meets with you and the other group leaders several times each week, he already has a good idea of what will be included in each status report. For this reason, what he really wants is information on which to base

the report he has to write to Karen. In fact, he has given the four groups a checklist with questions to guide them in filling out the status report. Bill likes concise reports that avoid unnecessary technical terms.

With this analysis of your reader in mind, you can start to plan a strategy for writing the status report. You should try to be as concise as possible. In fact, you should try to help Bill do his work; include not only a description of your group's progress but also a summary paragraph that he can use in his report to Karen.

If you are writing to a large group, or if for some other reason you cannot analyze your audience in detail, you might find it useful to classify your readers in general categories, such as technical readers, managers, and general readers. This classification process provides a basic guide to your readers' backgrounds, expectations, and needs. You can then modify this basic guide to accommodate the special needs of the particular audience.

Technical readers can range from experts to technicians. When you write to technical readers, remember their needs. The expert feels quite at home with technical vocabulary and formulas. You can get into the details of the technical subject right away, without sketching in the background; the expert already knows it. The technician—for instance, a medical technician—is likely to need a brief orientation to the subject; unlike the expert, the technician isn't always familiar with the theoretical background. The technician generally cares more about practical applications than about theory; he or she needs schematic diagrams, a parts list, and step-by-step instructions to apply to a specific task.

The manager is harder to define than the technical person, for the word *manager* describes what a person does more than what a person is or knows. A manager makes sure the operations of the organization are smooth and efficient. The manager of the procurement department at a manufacturing plant, for example, is responsible for seeing that raw materials are purchased and delivered on time so that production will not be interrupted. He or she is not interested in the technical details of how the parts work, except to know whether other parts can be substituted if the usual supplies are unavailable.

Upper-level managers are responsible for longer-range concerns. They have to foresee problems years ahead. Is a technology that is currently being used at the company becoming obsolete? What are the newest technologies? How expensive are they? How much of a disruption to the overall operations would be required in converting to a new technology? What other plans would have to be postponed or dropped altogether? When would the conversion start to pay for itself? What has been the experience of other companies that have adopted the new technologies? Executives are involved with these and dozens of other questions that go beyond the day-to-day managerial concerns.

All managers are primarily interested in the bottom line. They have to get a job done on schedule; they don't have time to study and admire a theory the way an expert does. Rather, managers have to juggle such constraints as money, data, and organizational priorities, and make logical and reasonable decisions quickly.

When you write to a manager, try to determine his or her technical background and then choose an appropriate vocabulary and sentence length. Regardless of the individual's background, focus on the practical information the manager will need to carry out the job.

Sometimes you will address the general reader—who has little or no background in your subject. For example, a nuclear scientist reading about economics is a general reader, as is a banker reading about new drugs used to treat arthritis. In writing to the general reader, avoid technical vocabulary and concepts and translate jargon, no matter how acceptable it might be in your own field, into standard English idiom. Use relatively short sentences when you are discussing subjects that might be confusing. Use analogies and examples. Sketch in any special background so that your reader can follow your discussion easily. Concentrate on the implications for the general reader. For instance, in discussing a new substance that removes graffiti from buildings, focus on its effectiveness and cost, not on its chemical composition.

Often you will be addressing more than one person. And in many cases, these different readers will represent two or even three of the basic categories. A common situation, for instance, is for a technical memo to be addressed to a technical reader, with copies going to several managers. Twenty years ago, this posed no problem, because most managers were technical people as well. Now, however, because of the knowledge explosion and the increasing complexity of business, managers might know very little about the technical aspects of the organization's product or service.

Further complicating things is the photocopy machine, which has made it possible to send a memo to 50 or 100 readers quickly and inexpensively. In the good old days of carbon paper, a document would go to only three or four people, or it would have to be retyped. With the increasingly common electronic office, sending copies of everything to everyone will be even easier and less expensive.

If you think your document will go to a multiple audience, try to satisfy

the needs of your readers as much as is reasonably possible. If it is a memo to be read by a manager as well as by a technical person, plan to minimize the technical terminology and briefly define the necessary technical terms. If you are writing a longer document, make sure you have an abstract and an executive summary for the different kinds of readers who might not want to read the whole thing. Make sure you have a detailed table of contents to help all your readers find the information they seek.

Analyzing Your Purpose

Once you have analyzed your audience, think about your purpose: What do you want your audience to know or believe after they have finished reading the document?

Think in terms of verbs. Try to isolate a single verb that expresses your purpose, and keep it in mind as you plan the document. (Of course, a document might have several purposes; if that's the case, think of several verbs.) Here are a few examples of verbs that indicate typical purposes you might be trying to accomplish in technical documents. The list has been divided into two categories: verbs to use when you want to communicate information to your readers, and verbs to use when you want to convince them to accept a particular point of view.

"COMMUNICATING" VERBS	"CONVINCING" VERBS
to explain	to assess
to inform	to request
to illustrate	to propose
to review	to recommend
to outline	to forecast
to authorize	
to describe	

This classification is not absolute. For example, "to review" could in some cases be a "convincing" verb rather than a "communicating" verb: one writer's review of a complicated situation might be very different from another's review of the same situation.

Following are two examples of how you can use the verbs to create simple statements of purpose that will help you clarify your task:

1. This report describes the research project to determine the effectiveness of the new waste-treatment filter.

2. This memo proposes that we study new ways to distribute specification revisions to our sales staff.

As you devise your purpose statement, remember that your real purpose might differ from the purpose you express. For instance, if your real purpose is to persuade your reader to lease a new computer system rather than purchase, you might phrase the purpose this way: "to explain the advantages of leasing over purchasing." Many readers don't want to be "persuaded"; they want to "learn the facts."

Once you have analyzed your audience and your purpose, it is time to begin your research.

Doing Research

Doing research actually consists of two basic tasks: brainstorming and researching itself. You need to know what information you will need, and you need to get your hands on it.

If you already know quite a bit about your topic, you will probably prefer to brainstorm first, then do research on the subjects you want to include. If you know little about your topic, you might want to do some research first to better understand the subject. Keep in mind that the writing process is not linear: you do not proceed from one step to the next without looking back. Often, you will return to an earlier step to revise what you have already written.

Brainstorming

Brainstorming is a process used either by one person or by a group of people to generate a list of subjects that might belong in a document.

Take a piece of paper (or sit in front of a computer screen), and jot down a brief phrase to help you remember any point that you think should be in the document. Don't write out the sentence; it takes too much time. Don't evaluate ideas at this point; write them down even if they seem farfetched. You want to encourage, not discourage, ideas, whether you are brainstorming alone or in a group. You will probably skip around from one subject to another; that is what you're supposed to do. And don't worry if some of the phrases you jot down involve details of larger ideas embraced by other phrases. You can straighten them out later.

Brainstorming has one purpose: to free your mind so that you can think of as many items as possible that relate to your topic. When you start to construct your outline, you will find that some of the items in your brainstorming list do not belong in the final document. Just toss them out. The advantage of brainstorming is that it is the most efficient way to catalog those things that *might* be important to the document. Ironically, a more structured way to generate ideas would cause you to miss more of them.

Researching

Researching, the process of gathering data, involves such tasks as doing library research or data-base searches, gathering in-house documents, conducting interviews, and administering questionnaires. The nature of your subject and your understanding of your audience and purpose will determine what kinds of research—and how much research—you need.

A lab report might require nothing more than studying your lab notebook. A feasibility study of whether to purchase an expensive piece of equipment might involve your getting product specifications, studying impartial evaluations in trade magazines, and visiting the manufacturers as well as plants that have already purchased the equipment.

Your readers might require or prefer certain kinds of research. If an important reader believes strongly in the value of an in-house demonstration, you should try to arrange it even if the information that it will yield will be provided by another research method. Your purpose, too, affects your choice of research methods. If you want to persuade your readers to accept an unpopular point of view, you will need more research than you would if the point of view were popular.

Organizing

Once you have gathered your information, you have to decide on a pattern to use in developing your ideas into an outline. The analysis of your audience and your purpose is your best guide. At some point in your outline, you might decide that a chronological approach would best suit your purposes. At another point, a spatial pattern might seem best. There is no single approach to developing ideas.

However, as you work on your outline, keep in mind that there are some standard patterns that usually work well in particular situations. Understanding these different patterns and how to combine them to meet your specific needs gives you more options as you put together your document.

Choosing Methods of Development

The following five patterns of development can be used effectively in technical writing:

1. *Chronological.* The chronological (time) pattern works well when you want to describe a process. If your readers have to follow your discussion in order to perform a task or to understand how something happens, chronology is a natural pattern to use. However, do not use the chronological pattern merely to reproduce your own thinking and actions. Keep the reader's needs in mind.
2. *Spatial.* In the spatial pattern, items are organized according to their physical relationships to one another. The spatial pattern is effective in structuring a description of a physical object, such as the controls on an instrument panel.
3. *General to specific.* The movement from general information to more specific information is a pattern that is used often, especially for sales literature and reports intended for a multiple audience. The general information enables the manager to understand the basics of the discussion without having to read all the details. The technical reader, too, appreciates the general information, which serves as an introductory overview.
4. *More important to less important.* This pattern is effective even in describing events or processes that would seem to call for a chronological pattern. For instance, suppose you were ready to write a report to a client after having performed an eight-step maintenance procedure on a piece of equipment. A chronological pattern focusing on what you did would answer the following question: "What did I do, and what did I find?" A more-important-to-less-important pattern, focusing first on the problem areas and then on the no-problem areas, would answer the following question: "What were the most important findings of the procedure?" Most readers would probably be more interested, first, in knowing *what* you found than in *how* you found it.
5. *Problems-methods-solution.* A basic pattern for outlining a complete report is to begin with the problem, then discuss the methods you used, and then finish with the results, conclusions, and recommendations. This pattern is used often with physical-research reports, because it mirrors the inductive method that lies at the heart of most lab reports.

Keep in mind that these five patterns are only suggestions. Your analysis of your audience and purpose might indicate that another pattern—or that several patterns—might be more effective. For example, you might be working within a problem-methods-solution pattern but use a more-important-to-less-important pattern within some of the subsections.

Outlining

The purpose of outlining is to create a plan to help you as you draft the document. Unless someone else will be working from your outline, don't worry about what it looks like. You can create a traditional outline with subheadings listed under headings, or you can create a "planet" with "satellites" radiating out from the center. Concentrate on making it useful for you. You want a set of headings that will let you concentrate on your subject when you start to draft. Once you have revised the draft, you want a set of headings that will help your readers find the information they want. Keep these two goals in mind as you outline.

Remember the basic principle of outlining. You are classifying information by assigning it to logical categories. The information in a heading should equal the sum of its subheadings. Every item contained in the heading should appear in the subheadings. For instance, the heading "American automobile manufacturers" would be illogical if its subheadings were "Ford," "Chrysler," and "General Motors." There are other American car makers. If you are not going to discuss every American car maker, the heading should be changed to something such as "major American car manufacturers." In addition, every subheading should be a part of the idea expressed in the heading. It is illogical to have a heading that reads "components of a personal computer" if the subheadings read "central processing unit," "disk drive," "keyboard," and "disks"; disks are not components of a personal computer.

Make sure, too, that items on the same level represent the same level of importance. A listing of baseball equipment should not include "gloves," "balls," "wooden bats," and "metal bats." Although there are wooden bats and metal bats, the other items in the listing are not subclassified. The final two items should be subheadings of "bats." Whenever possible, use parallel grammar to express items on the same level.

Drafting

A rough draft is a preliminary version of the final document. Some rough drafts are rougher than others. An experienced writer might be able to write a draft, fix a comma here and a word there, and have a perfectly serviceable document. On the other hand, every writer remembers the stubborn report that refused to make any sense at all after several drafts.

Many writers devise their own technique for drafting, but the key is to write quickly. Leave the revision to the next step.

If you are working on paper, write on one side only so that you can cut and paste later. (Many writers draft on note cards, one paragraph per card, to make it easy to shuffle the blocks of text.) Triple-space so that you can add material easily. When you start to draft, remember that there is no reason to begin at the beginning of the document. Most writers like to begin with a section from the middle of the document, usually a technical section they are comfortable with. Postponing the introductory elements of the document makes sense for another reason. Because the introductory elements are based on the body—and have to flow smoothly into the body—you have to know what the body will say before you can introduce it. Until you have written the body, you cannot be sure what it is going to say. Therefore, unless you want to begin with introductory elements merely as an exercise to gain a sense of the structure of the document, start with the body.

To make sure they keep writing as fast as they can, many writers force themselves to draft for a specified period, such as an hour, without stopping. Writing the rough draft is closer in spirit to brainstorming than to outlining. When you make up the outline, you are concentrating hard, trying to determine precise relationships. When you write the draft, you are just trying to turn your outline into paragraphs as fast as you can.

Your goal is to get beyond "writer's block," the mental paralysis that sometimes sets in when you look down at a blank piece of paper. Write *something*: it will be easy to revise later. Some writers are so insistent on keeping the rough draft flowing that they refuse to stop even when they can't figure out an item on the outline; they just skip to the next item that makes sense. Later, they go back and try to figure out the difficult item.

Revising

Almost all writers agree on one thing: you should not try to revise your draft right after you have finished writing it. Even though you will notice some minor problems, you will not be able to assess whether what you have written is clear, comprehensive, and accessible.

Set the document aside for as long as you can—at least a few hours,

preferably a day or two. Try to work on something else so you can forget what you wrote. Only in this way can you approach the document as your readers will.

How exactly do you revise a draft? Naturally, there is no single way. But the following technique works for many writers.

In your first pass through the document, concentrate on the largest issues: content and organization. Recreate your outline by writing all your headings in a list. Perhaps in writing the draft you omitted a heading from your original outline. Check to see if all the material is included. Having completed the first draft, you might now have a somewhat different perspective on the subject. In the haste of drafting you might have added material that now looks irrelevant. If so, mark it. You might want to move it or omit it altogether. Look at the headings to see that the sequence is clear and logical: remember that you are trying to meet your audience's needs. If you now think that a different organizational pattern will work better, make the changes.

Looking at the length of the text between headings will give you a sense of the emphasis you have given to the different topics in the document. If a relatively minor topic seems to be treated at great length, check the draft itself. The problem might be merely that you created more headings at that point than you did in treating some of the other topics. But if your treatment is in fact excessive, mark passages to condense.

Reread your draft also for accuracy. Double-check the figures and tables. Have you provided all the necessary data? Are the data correct? Check them against your notes.

Once you are satisfied that you have included the right information in the right order, revise for style. Have you used an appropriate level of vocabulary for your audience? Have you used consistent terminology throughout and provided a glossary—a list of definitions—if any of your readers will need it?

Are the sentences grammatically correct? Have you avoided awkward constructions? Are all the words spelled correctly? Some writers like to read their document from the end to the beginning—one sentence at a time. This process helps them focus on each individual sentence.

No matter how carefully you revise, you will not be able to see everything that your readers will. Therefore, get other people's comments before you make your final copy. Other readers will be able to make valuable suggestions on major concepts, such as organization and coverage, and on smaller issues, too.

EXERCISES

1. Choose a topic you are familiar with and interested in (such as some aspect of your academic requirements, a neighborhood concern, or some issue of public policy). After creating a hypothetical audience and purpose, brainstorm for about ten minutes, listing as many items as you can that might be relevant in a report on the topic. Finally, after analyzing your list, arrange the items in an outline, discarding irrelevant items and adding necessary ones that occur to you.

2. The following brainstorming list was created by a student interning as a chemist with a large chemical company.

The subject is trihalomethanes (THMs), a series of chemicals that pollute drinking water. The Environmental Protection Agency has ruled that within one year all municipalities must reduce the percentage of THMs in drinking water to less than 0.05 percent. The chemical company for which the writer works currently has four methods for removing THMs: chlorine dioxide, polymers, phosphates, and activated carbon. Each method has advantages and disadvantages.

Turn this brainstorming list into an outline for a report to management of the chemical company, recommending that the company create an aggressive marketing document that can be distributed to municipal governments. Where necessary, note information that has to be added to the outline.

Then, turn the brainstorming list into an outline for a report addressed to the municipal government of a town for which the activated-carbon method of reducing THMs is the most effective. Where necessary, note information that has to be added to the outline.

EPA guideline
activated carbon advantages and disadvantages
publicity through environmental environment groups on THM danger
regions of country where THM is biggest problem
what are THMs
how to determine the best method
15 states currently analyzing water for THMs
start-up costs for activated carbon high
polymers already used for other applications in water supplies
chlorine dioxide can boost other chemical above limits
chemical structures of common THMs
safety record of our company excellent
phosphates are somewhat dangerous to handle
activated carbon can be reused; therefore cheap to use

financial penalties for noncompliance
relationship between temperature and THM incidence
describe our free water-analysis program
schedule for other states to start analyzing water
no product development or testing cost to us: 4 methods EPA okayed
chlorine dioxide working already in one plant
chlorine dioxide an effective broad-range biocide
phosphates also inhibit future THM growth
activated carbon requires little operator attention

3. The following excerpts from outlines contain logical flaws. Revise the excerpts to eliminate the flaws.

a. I. Advantages of college football
  A. Fosters school spirit
  B. Teaches sportsmanship
 II. Increases revenue for college
  A. From alumni gifts
  B. From media coverage

b. A. Types of Health-Care Facilities
  1. Hospitals
  2. Nursing homes
  3. Care at home
  4. Hospices

3

Technical Writing Style

The best technical writing style is inconspicuous; your readers should not be aware of your presence as a writer. They should not notice that you have a great vocabulary or that your sentences flow beautifully—even if these things are true. They should be aware only of the information you are conveying. The reason for this is simple. People don't read technical writing to appreciate style. They read it because they want to know what you have to say.

The word *style*, as it is used in this chapter, refers to word choice, sentence construction, and paragraph structure.

Determining the Appropriate Stylistic Guidelines

The key to good technical writing is careful revision. However, there are some stylistic guidelines that you might be able to determine before you start to draft the document. Learning the "house style" that your organization follows will cut down the time needed for revision.

An organization's stylistic preferences might be defined explicitly in a company style guide. Or the organization might use an outside style manual, such as *The Chicago Manual of Style*. Sometimes, the stylistic preferences are implicit: no style manual exists, but over the years a set of unwritten guidelines has evolved. If this is the case in your organization, the best way to learn the house style is to study the documents in the files and ask more-experienced coworkers for advice.

The following discussion covers three important stylistic matters about which organizations commonly have clear preferences:

1. active and passive voice
2. first, second, and third person
3. headings and lists

As discussed in Appendix C, a word processor is a valuable tool in making stylistic revisions. A number of style programs exist that can help you deal with some of the topics discussed here. Even without specialized programs, however, you can perform most of the same techniques with the search function of any word-processing program. Where appropriate, these techniques will be mentioned in this chapter.

Active and Passive Voice

There are two voices: active and passive. In the active voice, the subject of the sentence performs the action expressed by the verb. In the passive voice, the subject receives the action. (In the following example, the subject is italicized.)

ACTIVE

Many *physicists* support the big-bang theory.

PASSIVE

The big-bang *theory* is supported by many physicists.

In most cases, the active voice is preferable to the passive voice. The active-voice sentence more clearly emphasizes the actor. In addition, the active-voice sentence is shorter, because it does not require a form of the verb *to be* and the past participle, as the passive-voice sentence does. In the example, for instance, the verb is "support," rather than "is supported," and "by" is unnecessary.

The passive voice, however, is generally more appropriate in four cases:

1. The actor is clear from the context.

EXAMPLE

Students are required to take both writing courses.

The context makes it clear that the college requires that students take both writing courses.

2. The actor is unknown.

EXAMPLE

The comet was first referred to in an ancient Egyptian text.

We don't know *who* referred to the comet.

3. The actor is less important than the action.

EXAMPLE

The documents were hand-delivered this morning.

It doesn't matter *who* the messenger was.

4. A reference to the actor is embarrassing, dangerous, or in some other way inappropriate.

EXAMPLE

Incorrect data were recorded for the flow rate.

It might be inappropriate to say *who* recorded the incorrect data.

Many people who otherwise take little interest in grammar have strong feelings about the relative merits of active and passive. A generation ago, students were taught that the active voice is inappropriate because it emphasizes the person who does the work rather than the work itself and thus robs the writing of objectivity. In many cases, this idea is valid. Why write, "I analyzed the sample for traces of iodine," when you can say, "The sample was analyzed for traces of iodine"? If there is no ambiguity about who did the analysis, or if it is not necessary to identify who did the analysis, a focus on the action being performed is appropriate.

Supporters of the active voice argue that the passive voice creates a double ambiguity. When you write, "The sample was analyzed for traces of iodine," your reader is not quite sure who did the analysis (you or someone else) or when it was done (as part of the project being described or some time previously). Even though a passive-voice sentence can contain all the information found in its active-voice counterpart, often the writer omits the actor.

The best approach to the active-passive problem is to recognize how the two voices differ and use them appropriately.

For example, don't use both active and passive voices in the same sentence unless you have a good reason.

AWKWARD

The new catalyst produced good-quality foam, and a flatter mold was caused by the new chute-opening size.

BETTER

The new catalyst produced good-quality foam, and the new chute-opening size resulted in a flatter mold.

A number of computer programs on style can help you find the passive voice in your writing. With any word-processing program, however, you can search for *is* and *was*, the forms of the verb *to be* that are most commonly used in passive-voice expressions. In addition, searching for *-ed* and *-en* will isolate many of the past participles, which also appear in most passive-voice expressions.

First, Second, and Third Person

Closely related to the question of voice is the question of person. The term *person* refers to the different forms of the personal pronoun:

FIRST PERSON

I worked . . . , we worked . . .

SECOND PERSON

You worked . . .

THIRD PERSON

He worked . . . , she worked . . . , it worked . . . , the machine worked . . . , they worked . . .

Organizations that prefer the active voice generally encourage the use of the first-person pronouns: "We analyzed the rate of flow." Organizations that prefer the passive voice often *prohibit* the use of the first-person pronouns: "The rate of flow was analyzed."

Another question of person that often arises is whether to use the second or the third person in instructions. In some organizations, instructional material—step-by-step procedures—is written in the second person: "You begin by locating the ON/OFF switch." The second person is concise and easy to understand. Other organizations prefer the more formal third person: "The operator begins by locating the ON/OFF switch." Perhaps the most popular version is the second person in the imperative: "Begin by locating the ON/OFF switch." In the imperative, the *you* is implicit. Regardless of the preferred style, however, be consistent in your use of the personal pronoun.

Headings and Lists

Headings and lists are major stylistic features of reports, memos, and letters in many organizations.

HEADINGS

The main purpose of a heading is to announce the subject of the discussion that follows it: the word *Headings*, for example, tells you that this discussion will cover the subject of headings. For the writer, the use of headings eliminates the need to announce the subject in a sentence such as "Let us now turn to the subject of headings."

A second important purpose of a heading is to clarify for the reader the hierarchical relationships within the document. This text, for instance, is divided into 11 chapters, each of which is divided into major units. Many of the major units are subdivided again. Understanding the level of importance of the various discussions helps the reader concentrate on the discussion itself.

To signify different hierarchical levels in headings in typewritten manuscript, use capitalization, underlining, and indention. Capital (uppercase) letters are more emphatic than small (lowercase) letters; therefore, an all-capitals heading (PROCEDURE) signals a more important category of information than an initial-capital heading. Similarly, underlined headings are more emphatic than ones that aren't underlined. A heading that begins at the left margin is more emphatic than one that is indented several spaces. (Note, however, that some writers like to center their major headings.)

Many word processors offer additional formatting options, including boldface, italics, and different sizes of type. Boldface, of course, is more emphatic than regular type, just as large letters are more emphatic than small letters.

To further reinforce the hierarchical levels, you can add a numbering scheme, such as the traditional outline system or the decimal system (see Chapter 6, page 91).

When you make up the headings, keep in mind that you are not restricted to simple phrases. Your assessment of the writing situation might indicate that a heading such as "What Is Wrong with the Present System?" would be more effective than one such as "Problem." Although general terms such as *problem*, *results*, and *conclusions* are well known by engineers and scientists, general readers and upper-level managers might have an easier time with more-informative phrases or questions.

LISTS

Like headings, lists let you manipulate the placement of words on the page to improve the effectiveness of the communication.

Many sentences in technical writing are long and complicated:

We recommend that more work on heat-exchanger performance be done with a larger variety of different fuels at the same temperature, with similar fuels at different temperatures, and with special fuels such as diesel fuel and shale-oil-derived fuels.

The difficulty in this sentence is that the readers cannot concentrate on the information because they must worry about remembering all the *with* phrases following "done." If they could "see" how many phrases they had to remember, their job would be easier.

Revised as a list, the sentence is easier to follow.

We recommend that more work on heat-exchanger performance be done with:
1. a larger variety of different fuels at the same temperature
2. similar fuels at different temperatures
3. special fuels such as diesel fuels and shale-oil-derived fuels

In this version, the placement of the words on the page reinforces the meaning. The readers can easily see that the sentence contains three items in a series. And the fact that each item begins at the left margin helps, too.

Make sure the items in the list are presented in a parallel structure. (See Appendix A for a discussion of parallelism.)

NONPARALLEL

Here is the schedule we plan to follow:
1. construction of the preliminary proposal
2. do library research
3. interview with the Bemco vice-president
4. first draft
5. revision of the first draft
6. after we get your approval, typing of the final draft

PARALLEL

Here is the schedule we plan to follow:
1. write the preliminary proposal
2. do library research
3. interview the Bemco vice-president
4. write the first draft
5. revise the first draft
6. type the final draft, after we receive your approval

In this example, the original version of the list is sloppy, a mixture of noun phrases (items 1, 3, 4, and 5), a verb phrase (item 2), and a participial phrase preceded by a dependent clause (item 6). The revision uses parallel verb phrases and deemphasizes the dependent clause in item 6 by placing it after the verb phrase.

Note that reports, memos, and letters do not have to look "formal," with traditional sentences and paragraphs covering the whole page. Headings and lists make writing easier to read and understand.

Choosing the Right Words and Phrases

Choosing the right words and phrases requires constant, sustained concentration. Most writers don't worry much about word choice during the drafting stage, but they are always on the lookout while revising.

The following discussion includes eight basic guidelines for choosing the right word:

1. Be specific.
2. Avoid unnecessary jargon.
3. Avoid wordy phrases.
4. Avoid clichés.
5. Avoid pompous words.
6. Focus on the "real" subject.
7. Focus on the "real" verb.
8. Do not use sexist language.

Be Specific

Being specific involves using precise words, providing adequate detail, and avoiding ambiguity.

Wherever possible, use the most precise word you can. A Ford Tempo is an automobile, but it is also a vehicle, a machine, and a thing. In describing the Ford Tempo, the word *automobile* is better than *vehicle*, because the less specific word also refers to trains, hot-air balloons, and other means of transport. As the words become more abstract—from *machine* to *thing*, for instance—the chances for misunderstanding increase.

In addition to using the most precise words you can, be sure to provide enough detail. Remember that the reader knows less than you do. What might be perfectly clear to you might be too vague for the reader.

VAGUE

An engine on the plane experienced some difficulties.

What engine? What plane? What difficulties?

CLEAR

The left engine on the Jetson 411 unaccountably lost power during flight.

Avoid ambiguity. That is, don't let the reader wonder which of two meanings you are trying to convey.

AMBIGUOUS

After stirring by hand for ten seconds, add three drops of the iodine mixture to the solution.

After stirring the iodine mixture or the solution?

CLEAR

Stir the iodine mixture by hand for ten seconds. Then add three drops to the solution.

CLEAR

Stir the solution by hand for ten seconds. Then add three drops of the iodine mixture.

What should you do if you don't have the specific data? You have two options: to approximate—and clearly tell the reader you are doing so—or to explain why the specific data are unavailable and indicate when they will become available.

VAGUE

The leakage in the fuel system is much greater than we had anticipated.

CLEAR

The leakage in the fuel system is much greater than we had anticipated; we estimate it to be at least five gallons per minute, rather than two.

Several style programs isolate common vague terms and suggest more precise alternatives.

Avoid Unnecessary Jargon

Jargon is shoptalk. To a banker, *CD* means certificate of deposit; to an audiophile, *CD* is a compact disc. The term *CAFE* (Corporate Average Fuel Economy) is a meaningful acronym among auto executives. The word *peripheral* means one thing to a computer specialist, another to the general reader. Although jargon is often held up to ridicule, it is a useful and natural kind of communication in its proper sphere. Two baseball pitchers would find it hard to talk to each other about their craft if they couldn't use terms such as *slider* and *curve*.

In one sense, the abuse of jargon is a needless corruption of the language. The best current example of this phenomenon is the degree to which computer-science terminology has crept into everyday English. In offices, employees are frequently asked to provide "feedback" or told that the coffee machine is "down." To many people, such words seem dehumanizing; to others, they seem silly. What is a natural and clear expression to one person is often strange or confusing to another. Communication breaks down.

An additional danger of using jargon outside a limited professional circle is that it sounds condescending to many people, as if the writer is showing off—displaying a level of expertise that excludes most readers. While the readers are concentrating on how much they dislike the writer, they are missing the message.

If some readers are offended by unnecessary jargon, others are intimidated. They feel somehow inadequate or stupid because they do not know what the writer is talking about. When the writer casually tosses in jargon, many readers really *can't* understand what is being said.

If you are addressing a technically knowledgeable audience, feel free to use appropriate jargon. However, an audience that includes managers or the general public will probably have trouble with specialized vocabulary. If your document has separate sections for different audiences—as in the case of a technical report with an executive summary—use jargon accordingly. A glossary (list of definitions) is useful if you suspect that the technical sections will be read by the managers.

Avoid Wordy Phrases

Wordy phrases weaken technical writing by making it unnecessarily long. Sometimes writers deliberately choose phrases such as "demonstrates a tendency to" rather than "tends to." The long phrase rolls off the tongue easily and appears to carry the weight of scientific truth. But the humble "tends to" says the same thing—and says it better for doing so concisely. The sentence "We can do it" is the concise version of "We possess the capability to achieve it."

Some wordy phrases just pop into writers' minds. We are all so used to hearing *take into consideration* that we don't realize that *consider* gets us there faster. Replacing wordy phrases with concise ones is therefore more difficult than it might seem. Avoiding the temptation to write the long phrase is only half the answer. The other half is to try to root out the long phrase that has infiltrated the prose unnoticed.

Following are a wordy sentence and a concise translation.

WORDY

I am of the opinion that, in regard to profit achievement, the statistics pertaining to this month will appear to indicate an upward tendency.

CONCISE

I think this month's statistics will show an increase in profits.

A special kind of wordiness to watch out for is redundancy, as in *end result, any and all, each and every, completely eliminate*, and *very unique*. Be content to say something once. Use "The liquid is green," not "The liquid is green in color."

REDUNDANT

We initially began our investigative analysis with a sample that was spherical in shape and heavy in weight.

BETTER

We began our analysis with a heavy, spherical sample.

Following is a list of some of the most commonly used wordy phrases and their concise equivalents.

WORDY PHRASE	CONCISE PHRASE
a majority of	most
a number of	some, many
at an early date	soon
at the conclusion of	after, following
at the present time	now
at this point in time	now
based on the fact that	because
despite the fact that	although
due to the fact that	because
during the course of	during
during the time that	during, while
having the capability to	can
in connection with	about, concerning
in order to	to
in regard to	regarding, about
in the event that	if
in view of the fact that	because
it is often the case that	often
it is our opinion that	we think that
it is our understanding that	we understand that
it is our recommendation that	we recommend that
make reference to	refer to
of the opinion that	think that
on the grounds that	because
prior to	before
relative to	regarding, about
so as to	to
subsequent to	after
take into consideration	consider
until such time	until

Avoid Clichés

Good writers avoid clichés. Phrases such as "Go for it" and "Make my day" are amusing for a while—a very short while. Then they become tiresome. Eventually, they can lose their meaning altogether.

A second problem with clichés is that people frequently get them wrong. For instance, the phrase "I could care less" is often used when the writer means the opposite.

Following are a cliché-filled sentence and a translation into plain English.

TRITE

Afraid that we were between a rock and a hard place, we decided to throw caution to the winds with a grandstand play that would catch our competition with its pants down.

PLAIN

Afraid that we were in a grave situation, we decided on a move that would surprise our competition.

Avoid Pompous Words

Writers sometimes try to impress their readers by using pompous words, such as *initiate* for *begin*, *perform* for *do*, and *prioritize* for *rank*. When asked why they use big words where small ones will do, writers say that they want to make sure their readers know they have a strong vocabulary, that they are well educated.

In technical writing, plain talk is best. If you know what you're talking about, be direct and simple. Even if you're not so sure what you're talking about, say it plainly; big words won't fool anyone for more than a few seconds.

Following is a pompous sentence translated into plain English.

POMPOUS

The purchase of a minicomputer will enhance our record maintenance capabilities.

PLAIN

Buying a minicomputer will help us maintain our records.

Following is a list of some of the most commonly used fancy words and their plain equivalents:

FANCY WORD	PLAIN WORD
advise	tell
ascertain	learn, find out
attempt	try
commence	start, begin
demonstrate	show
employ	use
endeavor (verb)	try
eventuate	happen
evidence (verb)	show
finalize	end, settle, agree
furnish	provide, give
impact (verb)	affect
initiate	begin
manifest (verb)	show
parameter	variable, condition
perform	do
prioritize	rank
procure	get, buy
quantify	measure
terminate	end, stop
utilize	use

Several style programs isolate fancy words and expressions. Of course, using any word-processing program, you can search for those terms that you tend to use inappropriately.

In the long run, your readers will be impressed by your clarity and accuracy. Don't waste your time thinking up fancy words.

Focus on the "Real" Subject

The conceptual or "real" subject of the sentence should also be the sentence's grammatical subject, and it should appear prominently in technical writing. Don't bury the real subject of the sentence in a prepositional phrase following a useless or "limp" grammatical subject. In the following examples, notice how the limp subjects disguise the real subjects. (The grammatical subjects are italicized.)

WEAK

The *use* of this method would eliminate the problem of motor damage.

STRONG

This *method* would eliminate the problem of motor damage.

WEAK

The *presence* of a six-membered lactone ring was detected.

STRONG

A six-membered lactone *ring* was detected.

Another way to make the subject of the sentence prominent is to eliminate grammatical expletives. You can almost always remove expletive constructions—*it is . . .* , *there is . . .* , and *there are . . .*—without changing the meaning of the sentence.

WEAK

There are many problems that must be worked out.

STRONG

Many problems must be worked out.

WEAK

It is with great pleasure that I welcome you to our annual development seminar.

STRONG

With great pleasure I welcome you to our annual development seminar.

STRONG

I am pleased to welcome you to our annual development seminar.

Using the search function of any word-processing program, you can find most weak subjects: usually they're right after the word *of*. The various expletives, naturally, are also easy to find.

Focus on the "Real" Verb

A "real" verb, like a "real" subject, should be prominent in every sentence. Few stylistic problems weaken a sentence more than the nominalized verb: a verb that has been converted into a noun. A sentence with a nominalized verb needs another verb (usually it's a weaker one) to be complete. "To install" becomes "to effect an installation," "to analyze" becomes "to conduct an analysis." Note how nominalizing the real verbs makes the following sentences both awkward and unnecessarily long. (The nominalized verbs are italicized.)

WEAK

Each *preparation* of the solution is done twice.

STRONG

Each solution is prepared twice.

WEAK

An *investigation* of all possible alternatives was undertaken.

STRONG

All possible alternatives were investigated.

Watch out for the *-tion* endings. They generally signify a nominalized verb that can be eliminated with a little rewriting.

Some software programs search for the most common nominalizations. With any word-processing program you can catch most of the nominalizations if you search for character strings such as *-tion*, *-ment*, *-ence*, and *-ance*. Many nominalized verbs are used with the preposition *of*.

Nominalized verbs are often used in passive-voice constructions; therefore, as you search for the forms of the verb *to be* to eliminate unnecessary passives, you will frequently find unnecessary nominalizations.

Do Not Use Sexist Language

Sexist language favors one sex at the expense of the other. Although sexist language can shortchange men—as in writing about some female-dominated professions such as nursing—in most cases, women are victimized. One example is using nouns such as *workman* and *chairman* when referring to a woman. Another is using a masculine pronoun when referring to both sexes, as in the sentence, "Each worker is responsible for *his* work area."

Over the years, different organizations have created synthetic pronouns, such as *thon*, *tey*, and *hir*, but these have never caught on. However, many organizations have formulated guidelines in an attempt to eliminate sexist language.

The relatively simple first step is to eliminate the male-gender words when they refer to both sexes. *Chairman*, for instance, can be replaced by *chairperson* or *chair*. *Firemen* are *firefighters*; *policemen* are *police officers*.

Rewording a sentence to eliminate masculine pronouns is also effective.

SEXIST

The operator should make sure he logs in.

NONSEXIST

The operator should make sure to log in.

In this revision, an infinitive replaces the *he* clause.

NONSEXIST

Operators should make sure they log in.

In this revision, the masculine pronoun is eliminated through a switch from singular to plural.

Notice that sometimes the plural can be unclear:

> UNCLEAR
>
> **Operators are responsible for their operating manuals.**

Does each operator have one operating manual or more than one?

> CLEAR
>
> **Each operator is responsible for his or her operating manual.**

In this revision, "his or her" clarifies the meaning. *He or she* and *his or her* are awkward, if overused, but they are at least clear.

If you use a word processor, search for *he*, *man*, and *men*, the words and parts of words most commonly associated with sexist writing. Some style programs search out common sexist terms and suggest nonsexist alternatives.

Creating Effective Paragraphs

An effective paragraph is easy to understand. The reader should know right away what the main point is, should understand how the body of the paragraph supports that main point, and should know how the paragraph relates to the material before and after it. The following discussion covers paragraph structure, length, and coherence.

Paragraph Structure

Start with the main point—the topic sentence. Technical writing should be clear and easy to read, not full of suspense. If a paragraph describes a test you performed on a piece of equipment, include the result in your first sentence: "The point-to-point continuity test on Cabinet 3 revealed no problems." Then go on to explain the details. If the paragraph describes a complicated idea, start with an overview: "Mitosis occurs in five stages: (1) interphase, (2) prophase, (3) metaphase, (4) anaphase, and (5) telophase." Then describe each stage. In other words, put the "bottom line" on top.

Note, for instance, how difficult the following paragraph is, because the

writer structured the discussion in the same order she performed her calculations:

> **Our estimates are based on our generating power during eight months of the year and purchasing it the other four. Based on the 1985 purchased power rate of $0.034/KW (January through April cost data) inflating at 8 percent annually, and a constant coal cost of $45–$50, the projected 1988 savings resulting from a conversion to coal would be $225,000.**

Putting the bottom line on top makes the paragraph much easier to read. Note how the writer adds a numbered list after the topic sentence.

> **The projected 1988 savings resulting from a conversion to coal are $225,000. This estimate is based on three assumptions: (1) that we will be generating power during eight months of the year and purchasing it the other four, (2) that power rates inflate at 8 percent from the 1985 figure of $0.034/KW (January through April cost data), and (3) that coal costs remain constant at $45–$50.**

The topic sentence in technical writing functions just as it does in any other kind of writing: it summarizes or forecasts the main point of the paragraph.

After the topic sentence comes the support. The purpose of the support is to make the topic sentence clear and convincing. Sometimes a few explanatory details can provide all the support needed. In the paragraph about estimated fuel savings, for example, the writer simply fills in the assumptions she used in making her calculation: the current energy rates, the inflation rate, and so forth. Sometimes, however, the support must carry a heavier load: it has to clarify a difficult thought or defend a controversial one.

Because every paragraph is unique, it is impossible to define the exact function of the support. In general, however, the support fulfills one of the following roles:

1. to define a key term or idea included in the topic sentence
2. to provide examples or illustrations of the situation described in the topic sentence
3. to identify factors that led to the situation described in the topic sentence
4. to define implications of the situation described in the topic sentence
5. to defend the assertion made in the topic sentence

The techniques used in developing the support include those used in most nonfiction writing: definition, comparison and contrast, classification and partition, and causal analysis.

Paragraph Length

How long should a paragraph of technical writing be? In general, 75 to 125 words will provide enough space for a topic sentence and four or five supporting sentences. Long paragraphs are more difficult to read than short paragraphs, for the simple reason that the readers have to concentrate longer. Long paragraphs also intimidate many readers by presenting long, unbroken stretches of type. Some readers actually will skip over long paragraphs.

Don't let an arbitrary guideline about length take precedence over your analysis of the writing situation. Often, you will need to write very brief paragraphs. You might need only one or two sentences—to introduce a graphic aid, for example. A transitional paragraph—one that links two other paragraphs—also is likely to be quite short. If a brief paragraph fulfills its function, let it be. Do not combine two ideas in one paragraph in order to achieve a minimum word count.

While it is confusing to include more than one basic idea in a paragraph, the concept of unity can be violated in the other direction. Often, you will find it necessary to divide one idea into two or more paragraphs. A complex idea that would require 200 or 300 words probably should not be squeezed into one paragraph.

Paragraph Coherence

After you have blocked out the main structure of the paragraph—the topic sentence and the support—make sure the paragraph is coherent. In a coherent paragraph, thoughts are linked together logically and clearly. Parallel ideas are expressed in parallel grammatical constructions. If the paragraph moves smoothly from sentence to sentence, emphasize the coherence by adding transitional words and phrases, repeating key words, and using demonstratives.

Maintaining coherence *between* paragraphs is the same process as maintaining coherence *within* paragraphs: place the transitional device as close as possible to the beginning of the second element. For example, the link between two sentences within a paragraph should be near the start of the second sentence:

The new embossing machine was found to be defective. *However*, the warranty on the machine will cover replacement costs.

The link between the two paragraphs should be near the start of the second paragraph:

The complete system would be too expensive for us to purchase now.......
..
..

***In addition*, a more advanced system is expected on the market within six months ....**
..
..

Transitional words and phrases help the reader understand a discussion by pointing out the direction the thoughts are following. Here is a list of the most common logical relationships between two thoughts and some of the common transitions that express those relationships:

RELATIONSHIP	TRANSITIONS
addition	also, and, finally, first (second, etc.), furthermore, in addition, likewise, moreover, similarly
comparison	in the same way, likewise, similarly
contrast	although, but, however, in contrast, nevertheless, on the other hand, yet
illustration	for example, for instance, in other words, to illustrate
cause-effect	as a result, because, consequently, hence, so, therefore, thus
time or space	above, around, earlier, later, next, to the right (left, west, etc.), soon, then
summary or conclusion	at last, finally, in conclusion, to conclude, to summarize

In the following example, the first version contains no transitional words and phrases. Note how much clearer the second version is.

WEAK

Neurons are not the only kind of cell in the brain. Blood cells supply oxygen and nutrients.

IMPROVED

Neurons are not the only kind of cell in the brain. *For example,* blood cells supply oxygen and nutrients.

Repetition of key words—generally, nouns—helps the reader follow the discussion. Note in the following example how the first version can be confusing.

UNCLEAR

For months, the project leaders carefully planned their research. The cost of the work was estimated to be over $200,000. (What is the *work*, the planning or the research?)

CLEAR

For months, the project leaders carefully planned their *research.* The cost of the *research* was estimated to be over $200,000.

Out of a misguided desire to be "interesting," some writers keep changing their important terms. *Plankton becomes miniature seaweed*, then *the ocean's fast food*. Leave this kind of word game to television sportscasters; technical writing must be clear.

In addition to transitional words and phrases and repetition of key phrases, demonstratives—*this, that, these,* and *those*—can help the writer maintain the coherence of a discussion by linking ideas securely. Demonstratives should in almost all cases serve as adjectives rather than as pronouns. In the following example, note that a demonstrative pronoun by itself can be confusing.

UNCLEAR

New screening techniques are being developed to combat viral infections. These are the subject of a new research effort in California. (What is being studied in California, *new screening techniques* or *viral infections?*)

CLEAR

New screening techniques are being developed to combat viral infections. *These techniques* are the subject of a new research effort in California.

Even when the context is clear, a demonstrative pronoun used without a

noun refers the reader to an earlier idea and therefore interrupts the reader's progress.

INTERRUPTIVE

The law firm advised that the company initiate proceedings. This resulted in the company's search for a second legal opinion.

FLUID

The law firm advised that the company initiate proceedings. *This advice* resulted in the company's search for a second legal opinion.

Transitional words and phrases, repetition, and demonstratives cannot *give* your writing coherence: they can only help the reader to appreciate the coherence that already exists. Your job is, first, to make sure your writing is coherent and, second, to highlight that coherence.

EXERCISES

1. The following sentences are vague. Revise the sentences to substitute specific information for the vague elements. Make up any reasonable details.

   a. The results won't be available for a while.
   b. The fire in the laboratory caused extensive damage.

2. Consider the voice (active or passive) in the following sentences. Which aspect of each sentence is emphasized? In what way would changing each sentence to the other voice change the emphasis?

   a. The proposal was submitted to the Planning Commission by Dr. Hendrick.
   b. On Tuesday, employee John Pawley drove the one-millionth Plymouth Reliant off the assembly line.

3. In the following sentences, the real subjects are buried in prepositional phrases or obscured by expletives. Revise the sentences so that the real subjects appear prominently.

   a. It is on the basis of recent research that I recommend the newer system.
   b. The use of point-of-purchase video presentations has resulted in a dramatic increase in business.

4. In the following sentences, unnecessary nominalization has obscured the real verb. Revise the sentences to focus on the real verb.

a. Pollution constitutes a threat to the Wilson Wildlife Reserve.
b. The construction of each unit will be done by three men.

5. The following sentences contain wordy phrases. Revise the sentences to make them more direct.

a. As far as experimentation is concerned, much work has to be done on animals.
b. The analysis should require a period of three months.

6. The following sentences contain clichés. Revise the sentences to eliminate the clichés.

a. With our backs to the wall, we decided to drop back and punt.
b. If we are to survive this difficult period, we are going to have to keep our ears to the ground and noses to the grindstone.

7. The following sentences contain pompous words. Revise the sentences to eliminate the pomposity.

a. This state-of-the-art beverage procurement module is to be utilized by the personnel associated with the Marketing Department.
b. It is indeed a not unsupportable inference that we have been unsuccessful in our attempt to forward the proposal to the proper agency in advance of the mandated date by which such proposals must be in receipt.

8. The following sentences contain sexist language. Revise the sentences to eliminate the sexism.

a. Each nurse is asked to make sure she follows the standard procedure for handling patient-information forms.
b. Policemen are required to live in the city in which they work.

9. In each of the following compound sentences, the coordinating conjunction *and* links two clauses. Revise the sentences to make the link between the two clauses stronger.

a. We interviewed Patricia Karney, president of Whelk Industries, and she said that our suggestions seem very interesting.
b. This new schedule will give the employees more freedom, and they will be more productive.

10. The following sentences might be too long for some readers. Break each one into two or more sentences. If appropriate, add transitional words and phrases or other coherence devices.

a. In the event that we get the contract, we must be ready by June 1 with the necessary personnel and equipment to get the job done, so with this end in mind a staff meeting, which all group managers are expected to attend, is scheduled for February 12.
b. Once we get the results of the stress tests on the 125-Z fiberglass mix, we will have a better idea where we stand in terms of our time constraints, because if it isn't suitable we will really have to hurry to find and test a replacement by the Phase I deadline.

11. The following examples contain choppy and abrupt sentences. Combine sentences to create a smoother prose style.

a. I need a figure on the surrender value of a policy. The policy number is A6423146. Can you get me this figure by tomorrow?
b. There are advantages to having your tax return prepared by a professional. There are also disadvantages. One of the advantages is that it saves you time. One of the disadvantages is that it costs you money.

12. The information contained in each of the following items could be conveyed more effectively in a list. Rewrite each sentence in the form of a list.

a. The freezer system used now is inefficient in several ways: the chef cannot buy in bulk or take advantage of special sales, there is a high rate of spoilage because the temperature is not uniform, and the staff wastes time buying provisions every day.
b. The causes of burnout can be studied from three areas: physiological—the roles of sleep, diet, and physical fatigue; psychological—the roles of guilt, fear, jealousy, and frustration; environmental—the roles of the physical surroundings at home and at work.

13. Provide a topic sentence for the following paragraph.

____
____.

The reason for this difference is that a larger percentage of engineers working in small firms may be expected to hold high-level positions. In firms with fewer than twenty engineers, for example, the median income was $33,200. In firms of 20 to 200 engineers, the median income was $30,345. For the largest firms, the median was $27,600.

14. Develop the following topic sentences into full paragraphs.

a. Job candidates should not automatically choose the company that offers the highest salary.
b. Every college student should learn at least the fundamentals of computer science.
c. The one college course I most regret not having taken is ______ .

15. In this exercise, several paragraphs have been grouped together into one long paragraph. Re-create the separate paragraphs by marking where the breaks should appear.

BOOKS ON COMPACT DISCS

Compact discs, which have revolutionized the recorded-music industry, are about to do the same for the book-publishing industry. A number of reference books, such as trade directories and multivolume encyclopedias, are already available in compact-disc format. Trade publishers are now working out the legal issues involved in publishing their books in compact-disc format. How is the compact-disc technology applied to books? The heart of the system is the same as that used for recorded music. Any kind of information that can be digitized—converted to the numbers zero and one—can be transferred to the 4.7-inch-diameter compact discs. Words and pictures, of course, are digitized in the common personal computer. Instead of outputting the digitized information exclusively as sound, the new technology connects a compact-disc player to a computer. The information stored on the disc is then output as words and pictures on the screen and as sound emitted through a speaker. Compact discs offer several important advantages over traditional delivery systems for printed information. First, compact discs can hold a tremendous amount of information. A 100-volume encyclopedia could fit on a single disc. The space storage advantages are considerable. Second, compact discs offer the accessing ease of an on-line system. If the user wants information on subatomic particles, he or she simply types in the phrase, and the system finds every reference to the subject. On request, the citations or even the entries themselves can be printed out on paper. And third, information stored on compact discs can be updated much less expensively than paper information. The subscriber or purchaser simply receives an updated disc periodically.

16. In the following paragraph, transitional words and phrases have been removed. Add an appropriate transition in each blank space. Where necessary, add punctuation.

As you know, the current regulation requires the use of conduit for all cable extending more than 18″ from the cable tray to the piece of

equipment. ______conduit is becoming increasingly expensive: up 17 percent in the last year alone. ______we would like to determine whether the NRC would grant us any flexibility in its conduit regulations. Could we ______run cable without conduit for lengths up to 3′ in low-risk situations such as wall-mounted cable or low-traffic areas? We realize ______that conduit will always remain necessary in high-risk situations. The cable specifications for the Unit 2 report to the NRC are due in less than two months; ______we would appreciate a quick reply to our request, as this matter will seriously affect our materials budget.

4

Graphic Aids

Graphic aids are the pictures of technical writing: diagrams, charts, graphs, tables, and photographs. Graphic aids are used extensively in technical writing because in some communication contexts they offer several important advantages over traditional text.

First, they are interesting to look at. The average reader who picks up a technical document will stop at the graphic aids and study them. Second, they are easy to understand and remember. Try describing in words what a hammer looks like. You could do it, but it might take a while, and some of your readers might misunderstand you. In 10 seconds, however, you could draw a clear, memorable sketch. Third, graphic aids are much more economical than text in communicating some kinds of information. A table can list the populations of each of the states over the last 10 years; the same information in text would require many paragraphs and would be very difficult to access. In general, technical information is most effectively communicated in some combination of graphics and text.

An effective graphic aid is:

1. appropriate to your audience and purpose
2. clear and complete
3. placed in an appropriate location
4. integrated with your text

First, of course, you need to determine whether a graphic aid is more appropriate than ordinary text. Graphic aids are particularly useful in clarifying and reinforcing difficult concepts, showing physical relationships among components of mechanisms, and demonstrating statistical trends and relationships. But you shouldn't create a graphic aid just for its own sake. If you think a graphic aid might help your readers, ask yourself what kind would work best. If you are showing general readers how an organization divides its advertising expenditures among television, radio, and print media, a pie chart might be the best kind of graphic to use. A more technical audience might be able to handle complicated tables or flowcharts.

Make sure the graphic aid is clear and complete. The reader should be able to grasp all the information contained in the graphic aid. To make it clear, label all the axes and variables on a graph. Identify the units of measure you are using, and indicate the time frame your data cover. Label the columns of a table clearly and accurately. If you did not generate the information yourself, credit the source.

Place the graphic aid in the most appropriate location in the document. Place it right after your first reference to it in the body of the document if you think that will help your readers understand the point. If your purpose is merely to reinforce or elaborate a point that is already clear, the appendix might be the best location.

Be sure to explain the main point of the graphic aid. A graphic aid can convey facts, but it cannot explain what those facts mean. A table showing population trends in the different states over the last decade can yield literally thousands of interpretations. Don't expect your readers to infer the one point you wish to communicate. Instead of merely paraphrasing the title of the graphic aid—"Table 3 shows the populations of the 50 states over the last decade"—make your point directly. "Table 3 shows that while the population of the Northeast and Midwest has declined by more than 5 percent over the last decade, the Sunbelt states have gained a net 21 percent in population."

There are dozens of types of graphic aids. Graphic artists are constantly modifying the basic bar graphs and line graphs to add visual interest. However, many writers feel that the most effective graphic aids are the most common ones—basic tables and graphs, for instance—because readers understand them easily and are not distracted from the information they contain. The increasing popularity of graphics software has made it possible for people without any artistic talent to create these common graphic aids quickly and effectively.

Graphic aids are generally classified into two categories: tables and figures. A table is a list of data—usually numbers—arranged in columns and rows. A figure is everything else: graphs, charts, diagrams, photographs, and so forth. Usually, tables and figures are numbered separately. The first table is Table 1; the first figure is Figure 1. In documents that are divided into sections or chapters, graphics are often identified by the unit. For instance, Figure 4-3 in a book would be the third figure in Chapter 4.

Tables

Tables can present vast amounts of quantitative information. If you want to show the number of people employed in six basic industries in ten different countries, a table can present the information clearly and concisely.

The basic components of a table are shown in Figure 4-1.

Table 1

Title
(subtitle)

Stub Heading	Column Heading 1	Column Heading 2	Column Heading 3
Stub Category 1			
Item A . . .	data[a]	data	data
Item B . . .	data	data	data
Item C . . .	data	data[b]	data
Item D . . .	data	data	data
Stub Category 2			
Item E . . .	data	data	data[c]
Item F . . .	data	data	data
Item G . . .	data	data	data

Notes: [a]Footnote
[b]Footnote
[c]Footnote
Source:

Figure 4-1. Parts of a Table

As you construct a table, keep in mind the following guidelines.

1. Create an informative title, such as "Test Results of Adhesive Formulas X1, X2, and X3." Be sure to include the necessary details in the title. For instance, don't title the table "Test Results." What test results? A common error is to omit the time frame being covered: "Incidence of Unexplained Motor Failure." Over how many years?
2. Structure the table so that it is read vertically, not horizontally. If you want to show your readers that the sales of cassettes in the United States in the last year have equaled those of phonograph records, structure the table this way:

Table 6 Recorded Music Sales (in millions)

	Year			
Type of Format	19--	19--	19--	19--
Compact disc	0	0	2.3	19.5
Phonograph record	119.8	89.6	73.1	65.7
Cassette	43.4	54.7	61.9	64.7

If, on the other hand, you want to emphasize the sales of each of the three formats over the four years, put the years on the vertical axis and the types of formats on the horizontal axis.

3. Clearly indicate the units of measure you are using. If all the data use the same units, state the unit in the title: "Farm Size in the Midwestern States (in Acres)." If the different columns or rows use different units, indicate the unit in the column or row headings.
4. Provide footnotes (using letters, not numbers, to prevent confusion) for any items that need explanation.
5. Group the items in the stub (the left-most column) in some logical order: alphabetical, chronological, topical, and so forth. For example:

   Sunbelt States
   - Arizona..
   - California..
   - New Mexico..

   Snowbelt States
   - Connecticut..
   - New York..
   - Vermont..

   Note that the writer skipped a line between groups for clarity. If your list contains no convenient groups, skip a line every four or five items for ease of reading. Also note the use of leader dots, which lead the reader's eyes from the stub item to the data itself.
6. Line up the data clearly in the columns:

   3,146
     573
   4,308

Figure 4-2 is an example of an effective table.

Table 6 Test Results for Valves No. 1 and No. 2

Valve Readings	*Maximum Bypass Cv*	*Minimum Recirc. Flow (GPM)*	*Pilot Threads Exposed*	*Main ΔP at Rated Flow (psid)*
Valve #1				
Initial	43.1	955	+3	4.5
Final	43.1	955	+3	. . . .
Valve #2				
Initial	48.1	930	+3	4.5
Final	48.1	950	+2	. . . .

Figure 4-2. An Effective Table

Bar Graphs

The two basic kinds of graphs are bar graphs and line graphs. Bar graphs effectively show the relationship between two different quantities. For example, a bar graph can show clearly that as the mortgage rates decrease, more people buy homes. The bars can be drawn either horizontally or vertically. However, horizontal bars are generally used to compare items at a given moment (such as quantities of different products sold during a single year); vertical bars are generally used to show how the quantity of an item changes over time (such as month-by-month sales of a single product).

Figure 4-3 shows the structure of a horizontal and vertical bar graph. When you construct bar graphs, keep in mind the following guidelines.

1. Make the proportions representative and easy-to-read. In a vertical bar graph, the vertical axis should be approximately three-quarters the length of the horizontal axis. In addition, make the space between the bars about one-half the width of a bar.
2. If possible, begin the quantity scale at zero. Note in Figure 4-4 that a quantity scale that does not begin at zero is misleading. If it is not practical to begin the quantity scale at zero, clearly break the axis, as shown in Figure 4-5.
3. If possible, label the exact quantities of the bars. Also, use tick marks or grid lines to indicate the quantities. Tick marks are the little marks drawn to the axis:

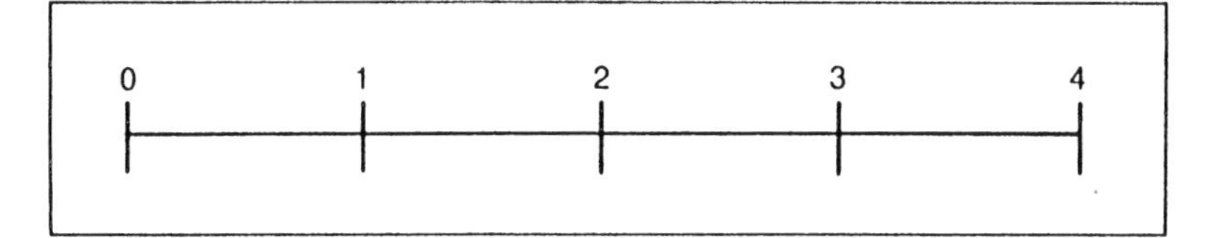

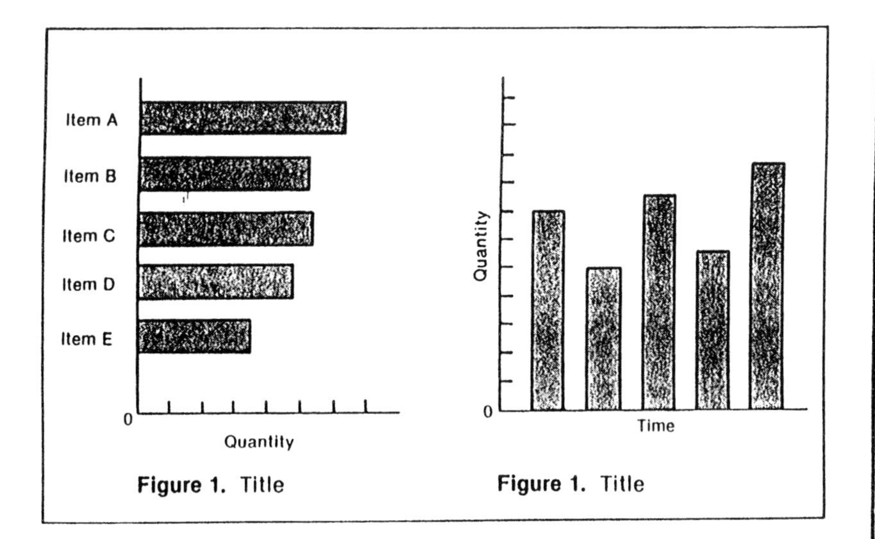

Figure 4-3 Structure of a Horizontal and a Vertical Bar Graph

Grid lines are ticks extended through the bars:

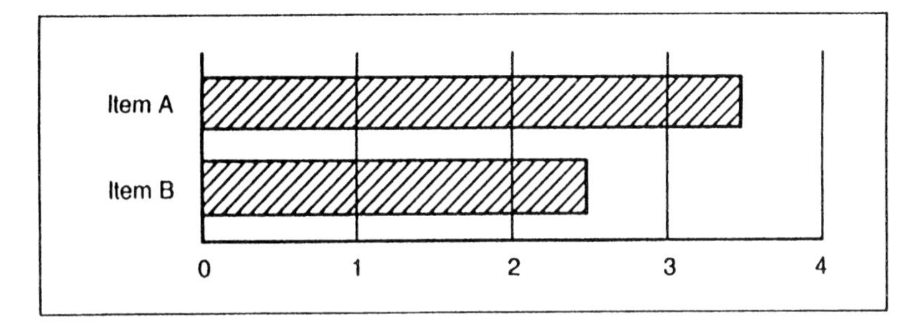

4. Arrange the bars in a logical sequence. In a vertical bar graph, chronology usually dictates the sequence. For a horizontal bar graph arrange the bars in descending-size order unless some other sequence seems more appropriate.

Figure 4-6 is an effective bar graph. Note that, like all figures, graphs are titled underneath. Tables are titled above.

The basic bar graph can be varied to accommodate many different communication needs. Following are a few common variations.

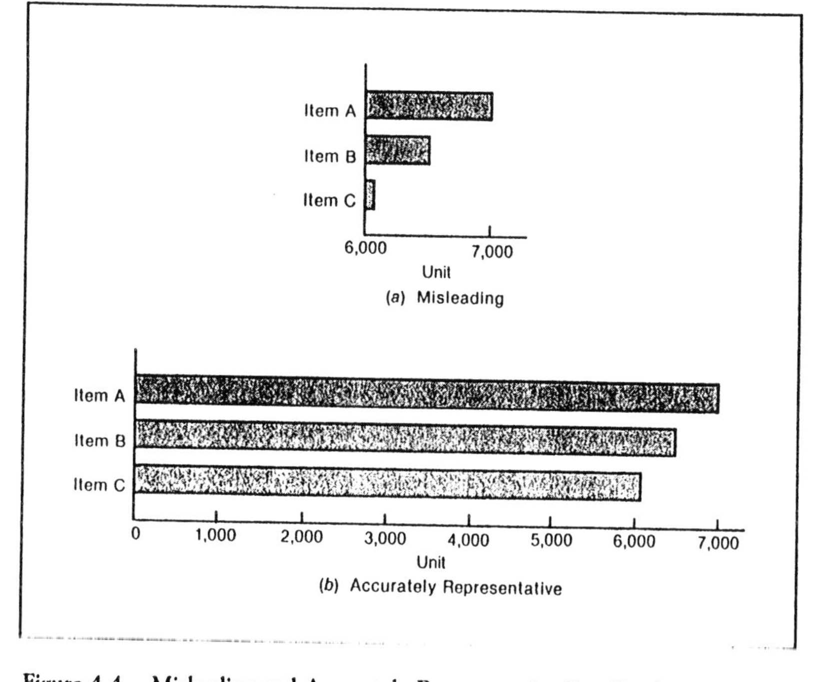

Figure 4-4. Misleading and Accurately Representative Bar Graphs

The *grouped bar graph*, such as that in Figure 4-7, lets you show several quantities for each item you are representing. Grouped bar graphs are useful for showing information such as the numbers of full-time and part-time students at several universities. One kind of bar represents the full-time students; the other, the part-time. To distinguish the bars from each other,

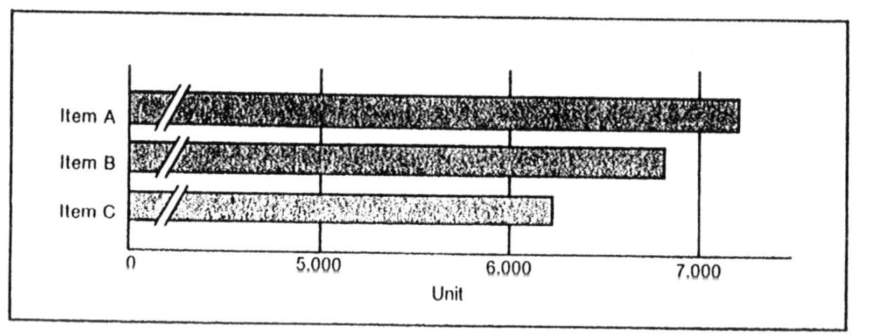

Figure 4-5. A Bar Graph with the Quantity Axis Clearly Broken

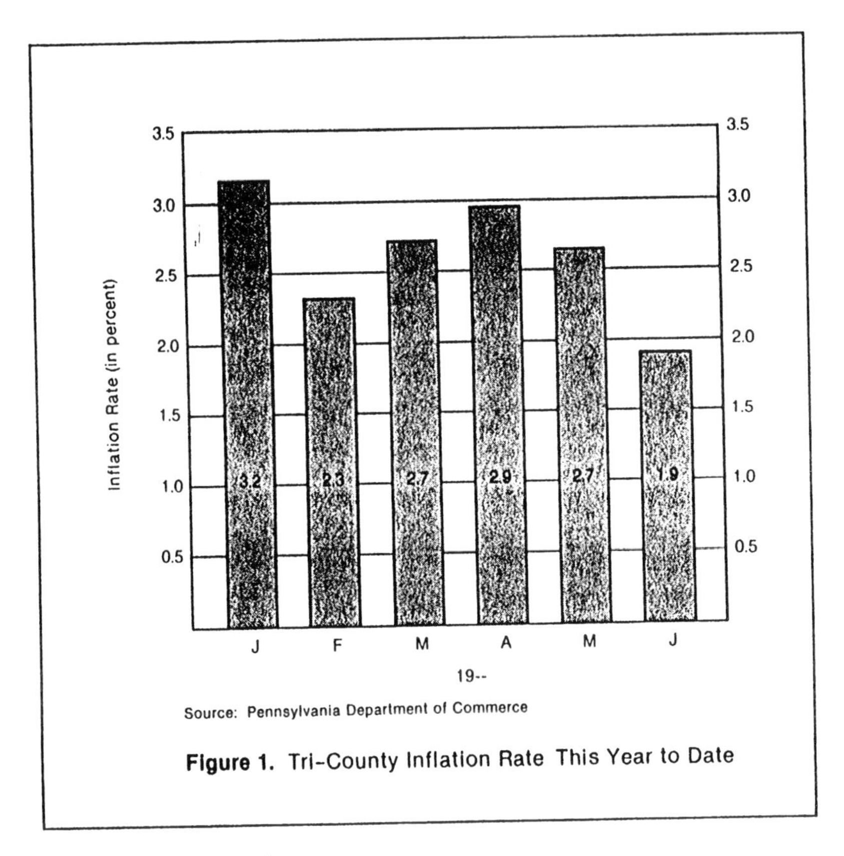

Figure 4-6. Bar Graph

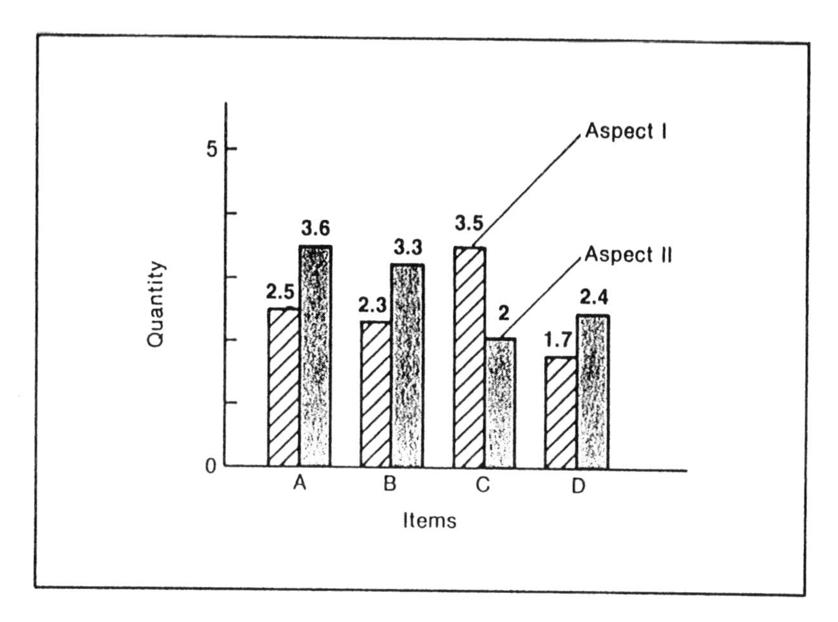

Figure 4-7. Grouped Bar Graph

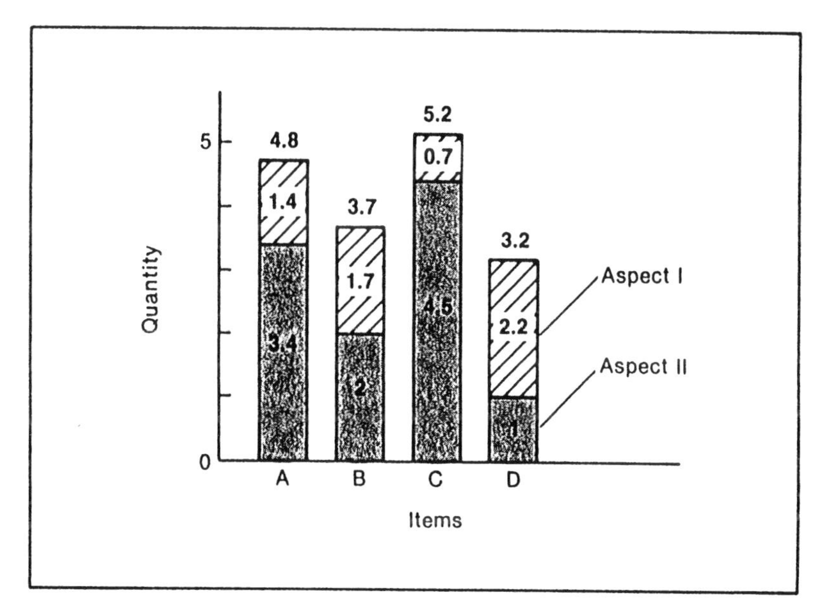

Figure 4-8. Subdivided Bar Graph

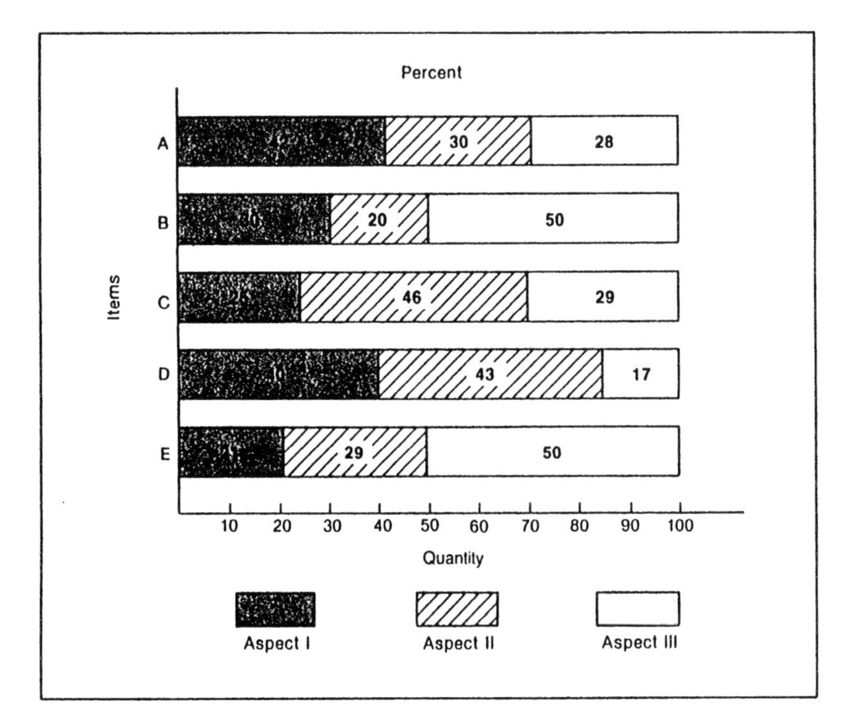

Figure 4-9. 100-Percent Bar Graph

use hatching (striping) or shading, and label one set of bars or provide a key. Leave at least one bar's width between sets of bars.

Another way to show this kind of information is through the *subdivided bar graph*, shown in Figure 4-8. A subdivided bar graph adds Aspect I to Aspect II, just as wooden blocks are placed on one another. Although the totals are easy to compare in a subdivided bar graph, the individual quantities (except those that begin on the horizontal axis) are not. Therefore, be sure to indicate the quantities.

Related to the subdivided bar graph is the *100-percent bar graph*, which enables you to show the relative proportions of the elements that make up several items. Figure 4-9 shows a 100-percent bar graph. This kind of graph is useful in portraying, for example, the proportion of full-scholarship, partial-scholarship, and no-scholarship students at a number of colleges.

The *deviation bar graph*, shown in Figure 4-10, lets you show how various quantities deviate from a norm. Deviation bar graphs are often used when the information contains both positive and negative values, as with profits

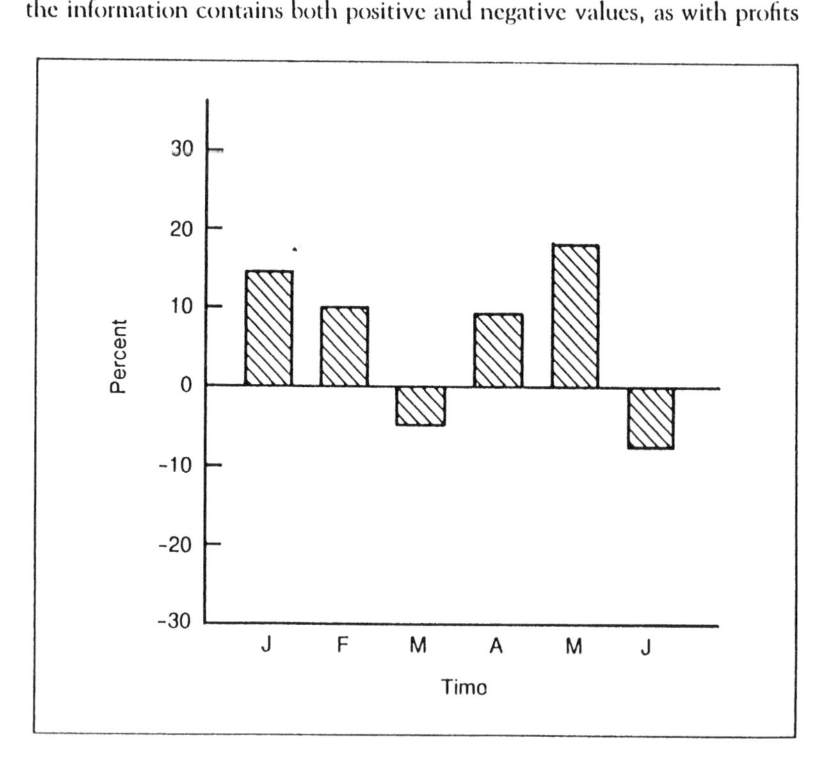

Figure 4-10. Deviation Bar Graph

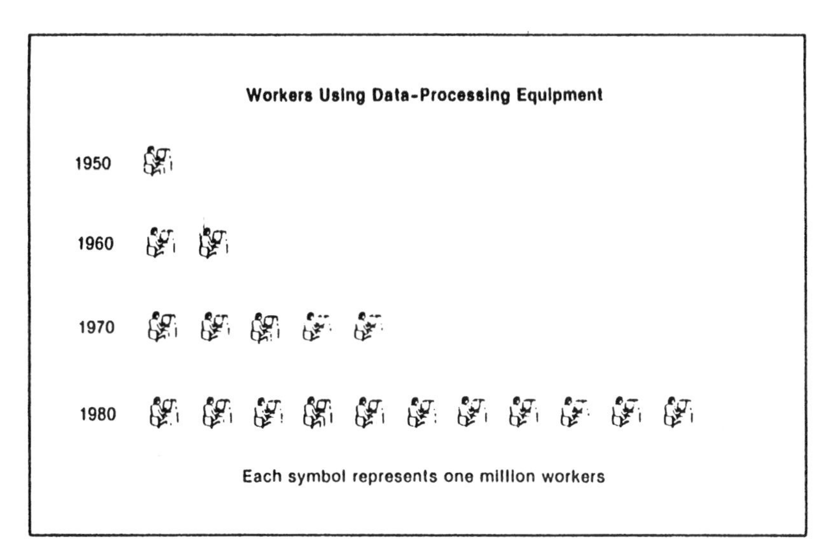

Figure 4-11. Pictograph

and losses. Bars on the positive side of the norm line represent profits; on the negative side, losses.

Pictographs are horizontal bar graphs in which the bars are replaced by a series of symbols that represent the items (see Figure 4-11). Pictographs are generally used only to enliven statistical information for the general reader. The quantity scale is usually replaced by a statement that indicates the numerical value of each symbol.

Line Graphs

Line graphs are like vertical bar graphs, except that in line graphs the quantities are represented not by bars but by points linked by a line. This line traces a pattern that in a bar graph would be formed by the highest point of each bar. Line graphs are used almost exclusively to show how the quantity of an item changes over time. Some typical applications of a line graph would be to portray the month-by-month sales figures or production figures for a product or the annual rainfall for a region over a given number of years. A line graph focuses the reader's attention on the change in quantity, whereas a bar graph emphasizes the quantities themselves.

An additional advantage of the line graph for demonstrating change is that it can accommodate much more data. Because three or four lines can

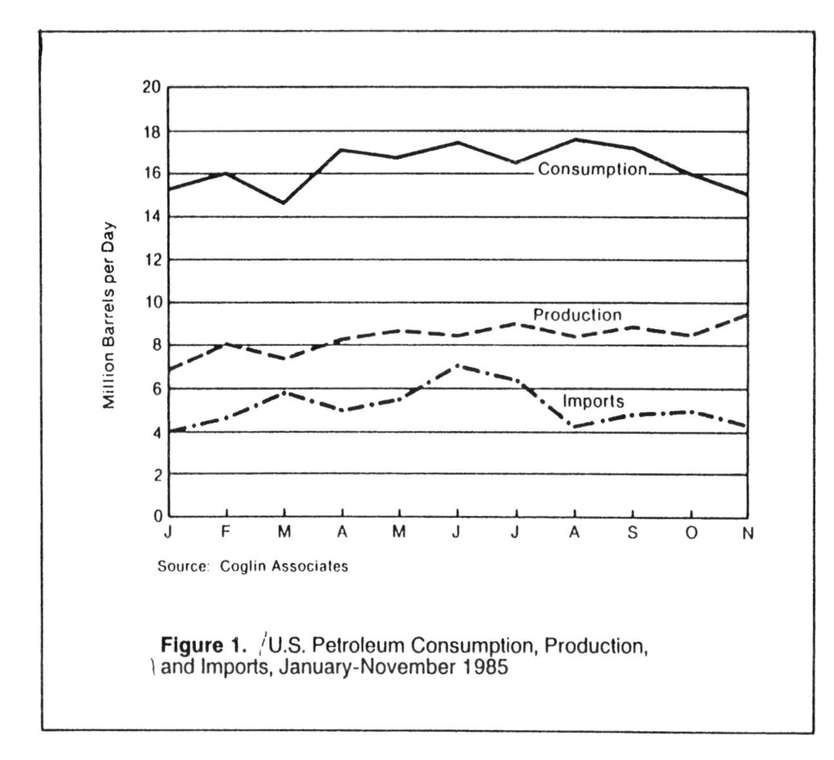

Figure 4-12. Multiple-line Graph

be plotted on the same graph, you can compare trends conveniently. Figure 4-12 shows a multiple-line graph. However, if the lines intersect each other often, the graph will be unclear. If this is the case, draw separate graphs.

The principles of constructing a line graph are similar to those used for a vertical bar graph. The vertical axis, which charts the quantity, should begin at zero; if it is impractical to begin at zero because of space restrictions, clearly indicate a break in the axis. When you need to be precise, use grid lines—horizontal, vertical, or both—rather than tick marks.

Charts

Whereas tables and graphs present statistical information, most charts convey more abstract relationships, such as causality or hierarchy. (The pie chart, which is really just a circular representation of the 100-percent bar graph, is the major exception.) Many forms of tables and graphs are well known and fairly standard. By contrast, only a few kinds of charts—such as the organization chart and flowchart—follow established patterns. Most charts reflect original concepts and are created to meet specific communication needs.

The *pie chart* is a simple but limited design used for showing the relative size of the parts of a whole. Pie charts can be instantly recognized and understood by the untrained reader: everyone remembers the perennial "where-your-tax-dollar-goes" pie chart. The circular design is effective in showing the relative size of as many as five or six parts of the whole, but it cannot easily handle more parts because, as the slices get smaller, judging their sizes becomes more difficult. (Very small quantities that would make a pie chart unclear can be grouped under the heading "Miscellaneous" and explained in a footnote.)

To create a pie chart, begin with the largest slice at the top of the pie and work clockwise in decreasing-size order, unless you have a good reason for arranging the slices in a different order. (If you have a miscellaneous section, place it last.) Label the slices (horizontally, not radially) inside the slice, if space permits. It is customary to include the percentage that each slice represents. Sometimes, the absolute quantity is added. To emphasize one of the slices—for example, to introduce a discussion of the item represented by that slice—separate it from the pie. Make sure your math is accurate as you convert percentages into degrees in dividing the circle. A percentage circle guide—a template with the circle already converted into percentages—is a useful tool.

Figure 4-13 shows two styles of pie charts. The chart on the left is the standard version with shading added to simulate a third dimension. The version on the right shows a slice removed from the pie for emphasis.

A *flowchart*, as its name suggests, traces the stages of a procedure or a process. A flowchart might be used, for example, to show the steps involved in transforming lumber into paper or in synthesizing an antibody. Flowcharts are useful, too, for summarizing instructions that a reader is to carry out. The basic flowchart portrays stages with labeled rectangles or circles. To make it visually more interesting, use pictorial symbols instead of geometric shapes. If the process involves quantities (for example, the process of paper manufacturing might "waste" 30 percent of the lumber), they can be listed or merely suggested by the size of the line used to connect the stages. Flowcharts can portray open systems (those that have a "start" and a "finish") or closed systems (those that end where they began). A special kind of flowchart, called a decision chart (in which the flow follows different routes depending on yes/no answers to questions), is used frequently in computer science.

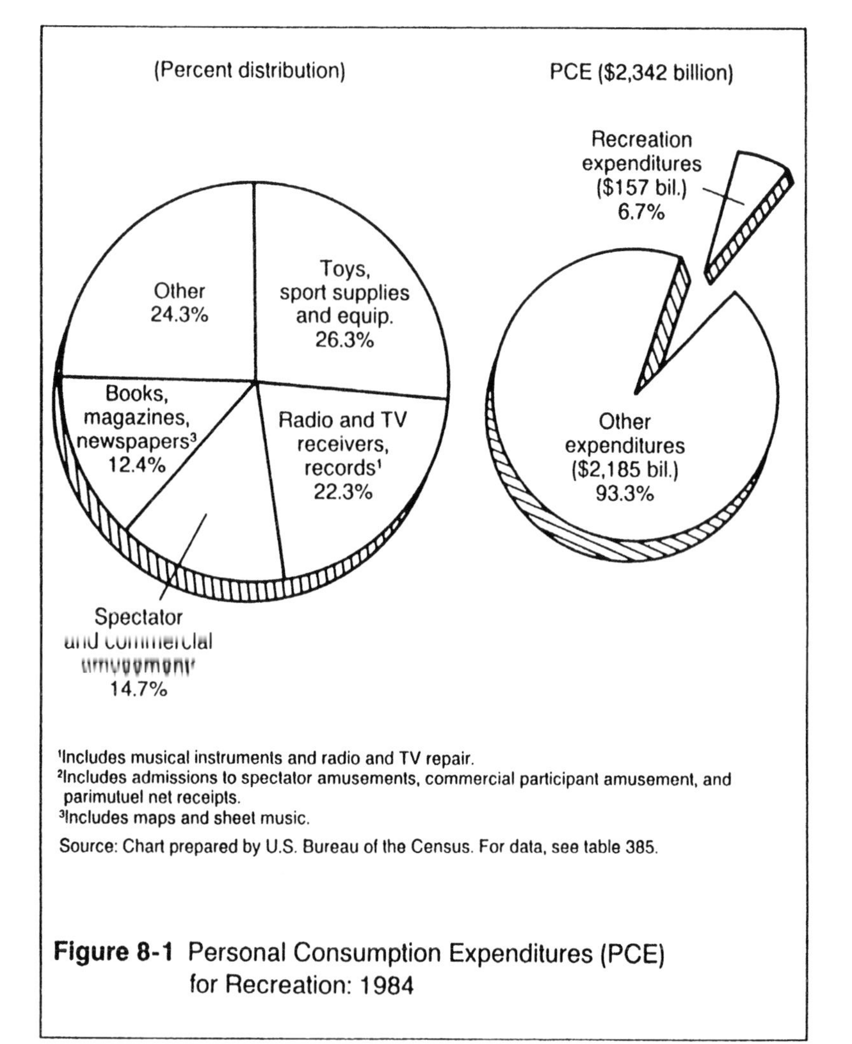

Figure 4-13. Pie Charts

Source: *Statistical Abstract of the United States,* 1986, p. 222

Figure 4-14 shows an open-system flowchart whose bars correspond in width to the magnitude of the quantity they represent. The subject of the flowchart is the percentage of solar energy that reaches the earth's surface. Figure 4-15 shows a typical flowchart from computer science. This chart incorporates the concept of the decision tree. The diamond-shaped stages of

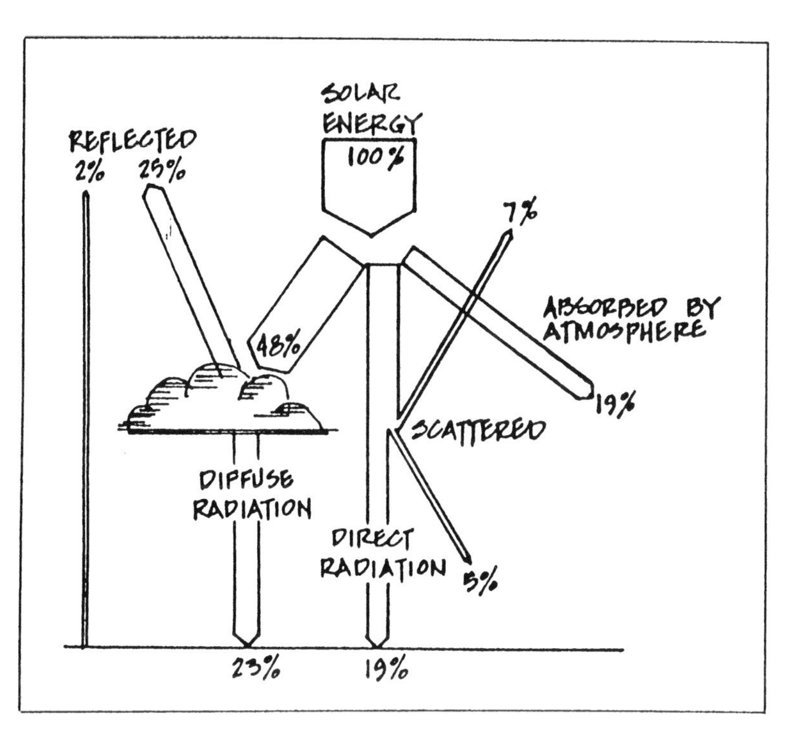

Figure 4-14. Open-System Flowchart

the process are questions that can be answered either yes or no. The process branches off at each diamond.

An *organization chart* is a type of flowchart that portrays the flow of authority and responsibility in a structured organization. In most cases, the positions are represented by rectangles. The more important positions can be emphasized through the size of the boxes, the width of the lines that form the boxes, the typeface, or the use of color. If space permits, the boxes themselves can include brief descriptions of the positions, duties, or responsibilities.

Figure 4-16 is a typical organization chart. Unlike other figures, organization charts are generally titled *above* the chart.

Diagrams and Photographs

In portraying physical relationships, such as those in pieces of equipment or machinery, diagrams and photographs are often the most effective type of graphic aids. Photographs are unmatched, of course, for reproducing

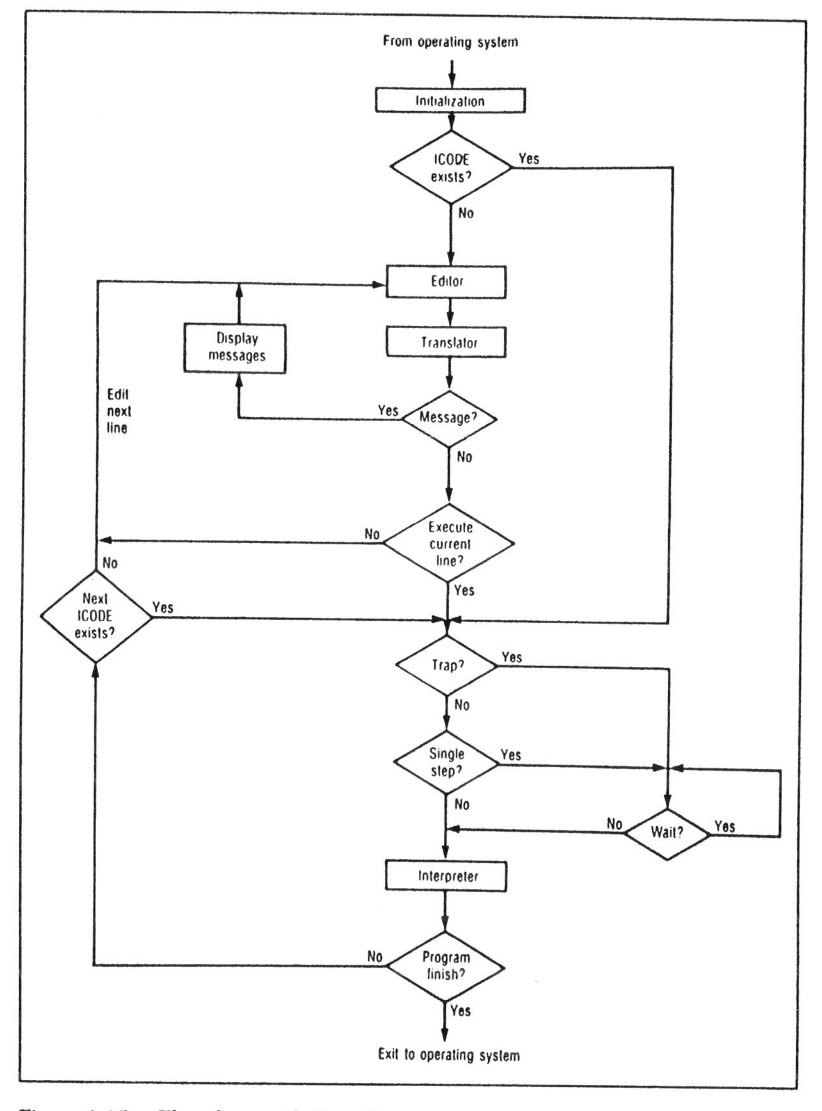

Figure 4-15. Flowchart with Branches

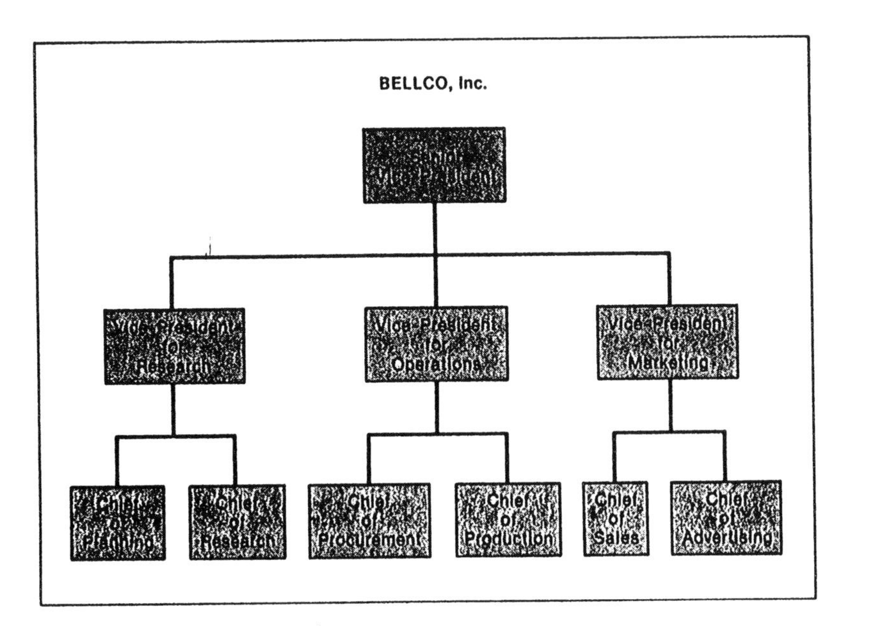

Figure 4-16. Organization Chart

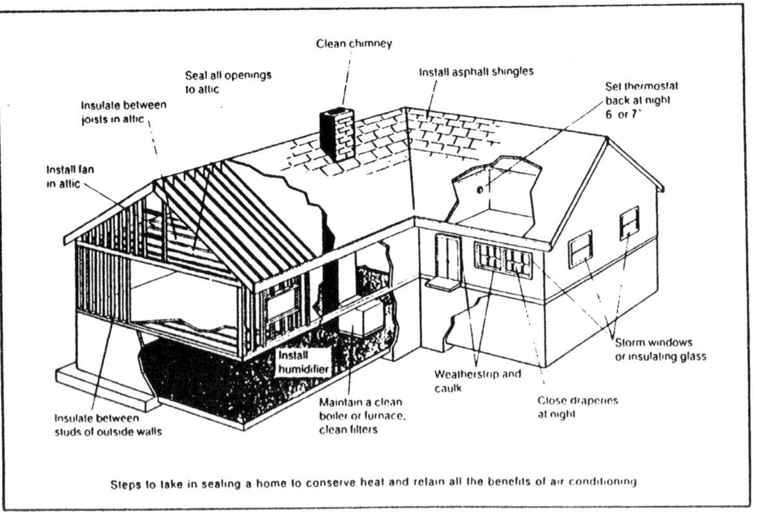

Figure 4-17. Cutaway Diagram

Figure 4-18. Exploded Diagram

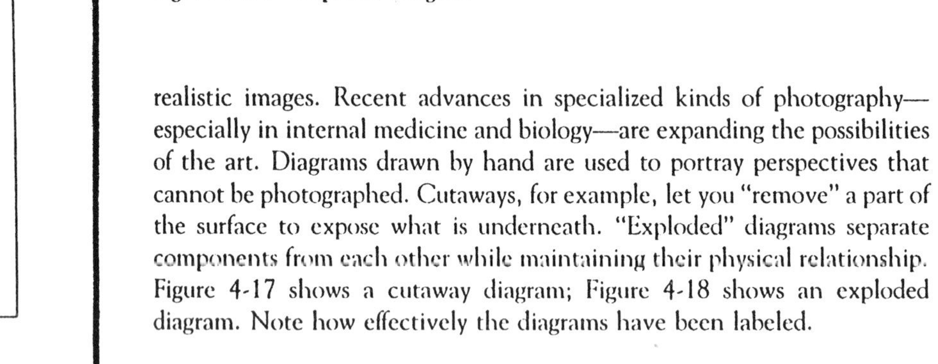

realistic images. Recent advances in specialized kinds of photography—especially in internal medicine and biology—are expanding the possibilities of the art. Diagrams drawn by hand are used to portray perspectives that cannot be photographed. Cutaways, for example, let you "remove" a part of the surface to expose what is underneath. "Exploded" diagrams separate components from each other while maintaining their physical relationship. Figure 4-17 shows a cutaway diagram; Figure 4-18 shows an exploded diagram. Note how effectively the diagrams have been labeled.

EXERCISES

1. In each of the following exercises, translate the information into at least two different kinds of graphic aids. For each exercise, which kind works best? If one kind works well for one audience but not so well for another audience, be prepared to explain.

   a. The prime interest rate had a major effect on our sales. In January, the rate was 11.5 percent. It went up a full point in February, and another half point in March. In April, it leveled off, and it dropped two full points each in May and June. Our sales figures were as follows for the Crusader 1: January, 5,700; February, 4,900; March, 4,650; April, 4,720; May, 6,200; June, 8,425.

   b. Following is a list of our new products, showing for each the profit on the suggested retail price, the factory where produced, the date of introduction, and the suggested retail price.

THE TIMBERLINE	THE FOUR SEASON
Profit 28%	Profit 32%
Milwaukee	Milwaukee
March 1984	October 1983
$235.00	$185.00
THE FAMILY EXCURSION	**THE DAY TRIPPER**
Profit 19%	Profit 17%
Brooklyn	Brooklyn
October 1983	May 1983
$165.00	$135.00

   c. In January of this year, we sold 50,000 units of the BG-1, of which 20,000 were purchased by the army. In February, the army purchased 15,000 of our 60,000 units sold. In March, it purchased 12,000 of the 65,000 we sold.

5

Letters and Memos

Some eighty million business letters are written every working day. No figures exist about memos, but presumably many more than that are written. As a professional, you can expect to write letters and memos often.

What is the difference between a letter and a memo? A letter is a short, written communication addressed in most cases to someone outside one's organization. A memo is addressed to someone within the organization. There are some format differences: the letter retains the traditional "Dear Mr. Hopkins" and "Very truly yours" elements that have been used for centuries, whereas the memo uses a more direct to-from-subject format.

A letter is usually more formal in tone than a memo, because in most cases the writer does not know the reader as well. If the writer has done business with the reader before, a cordial tone might be adopted. In some organizations, it is customary to refer to the reader by his or her first name.

One important similarity between letters and memos should be mentioned. Both are legally binding documents. A letter written on company stationery and a memo written on a preprinted memo pad are not personal statements; they can be used as evidence in court. Therefore, be sure that whatever you write in a letter or a memo reflects positively on you, your coworkers, and your organization. One way to think of this point is: If you are going to commit anything to paper, make sure you wouldn't mind if your organization's chief officer read it.

Letters

Like all technical writing, letters should be clear, accurate, comprehensive, accessible, concise, and correct. However, because letters are structured as person-to-person documents, keep in mind one other factor: the "you attitude."

The "You Attitude"

The "you attitude" is a courteous, positive tone. It involves looking at the situation from the reader's point of view. If, for example, a client orders a product that your organization no longer manufactures, it is impolite and therefore unwise to answer merely that you can't fill the order. Instead, put yourself in the reader's place. You should suggest a similar product that will meet the reader's needs, or if you don't manufacture a similar product, recommend another supplier. In other words, be helpful. The reader doesn't

care whether you can supply the product; he or she wants to solve a problem.

The "you attitude" extends to the tone you use. Don't focus on yourself. Focus on the reader. Following are several sentences that violate the "you attitude." After each sentence is an improved version.

POOR

You wrote to the wrong department. We don't handle complaints.

BETTER

Your letter has been forwarded to the Customer Service Department.

POOR

You must have dropped the engine. The engine is badly cracked.

BETTER

The badly cracked engine suggests that your engine must have fallen onto a hard surface from some height.

POOR

Only our award-winning research and development department could have devised this extraordinary sump pump.

BETTER

You will find that our new sump pump features significant advantages over ordinary sump pumps.

A calm, respectful, polite tone always makes the best impression and therefore increases your chances of achieving your goal.

Avoiding Letter Clichés

Related to the "you attitude" is the issue of letter clichés. Over the decades, a whole set of words and phrases have come to be associated with letters, such as "as per your request." For some reason, many people think that these phrases are required. They're not. In fact, they make a letter sound stilted and insincere. If you would feel awkward or uncomfortable saying these clichés to a friend, avoid them in your letters.

Following is a list of some of the common clichés and their more natural equivalents.

CLICHÉS	EQUIVALENTS
attached please find	attached is
at your earliest convenience	soon
cognizant of	aware that
enclosed please find	enclosed is
endeavor (verb)	try
herewith ("We herewith submit . . .")	*"Herewith" doesn't say anything. Skip it.*
hereinabove	previously, already
in receipt of ("We are in receipt of . . .")	"We have received . . ."
permit me to say	*Just say it.*
pursuant to our agreement	as we agreed
referring to your ("Referring to your letter of March 19, the shipment of pianos . . .")	"In reference to your letter of March 19, the . . .," *or subordinate the reference at the end of your sentence.*
same (*as a pronoun*: "Payment for same is requested.")	*Use the noun instead:* "Payment for the merchandise is requested."
wish to advise ("We wish to advise that . . .")	*The phrase doesn't say anything. Just say what you want to say.*
the writer ("The writer believes that . . .")	"I believe that . . ."

Elements of a Letter

Most letters have a heading, the date, an inside address, salutation, body, complimentary close, signature, and reference initials. In addition, some letters contain one or more of the following notations: "Attention," "Subject," "Enclosure," and "Copy."

HEADING

When you are writing on company stationery, the heading is the letterhead. When you are writing on other stationery, the heading consists of your address (but not your name). See the sample job-application letter, page 155, for an example of such a heading.

DATE

In expressing the date, do not use all numerals. For example, in the United States, "1/3/88" means January 3, 1988, but it means March 1, 1988, in most other countries. Use one of the two following forms:

January 3, 1988
3 January 1988

INSIDE ADDRESS

The inside address consists of the reader's name, position, organization, and address. Use courtesy titles such as *Dr.*, *Professor*, or *The Reverend*. Otherwise, use *Mr.* for men and *Ms.* for women (unless you know your reader prefers *Mrs.* or *Miss*).

Following is an example of an inside address:

Mr. John Pearsall, Director of Operations
The Hartford Supply Company
315 Winding Way
Hartford, CT 06413

SALUTATION

The salutation, which is placed two lines below the inside address, usually consists of the word *Dear* followed by the reader's courtesy title, last name, and a colon:

Dear Ms. Withers:

If you cannot determine the reader's name, use a general term, such as *Dear Technical Director*. To address groups, use general terms (*Dear Members of the Restoration Committee* or *Ladies and Gentlemen*).

BODY

The body is the substance of the letter. Generally, it will be three or more paragraphs. The first paragraph is usually a brief introduction. It announces the purpose of the letter and refers to the previous communication, as in the following example:

Thank you for your letter of March 15 requesting our latest parts catalog. We are pleased to send it to you.

The second and subsequent paragraphs elaborate the message. These paragraphs will differ in strategy, of course, according to your purpose. When you are communicating good news, the common strategy is to announce the good news immediately and then develop your main points. When you are communicating bad news, the common strategy is to withhold the bad news until you have had a chance to explain the circumstances or the reasons behind it. Otherwise, you might disappoint or anger your reader before you explain your side of the story.

The last paragraph concludes the letter. At this point, you summarize your message or describe any future actions that you or your reader will take regarding the subject discussed in the letter. You should try to sustain a positive, cooperative tone.

COMPLIMENTARY CLOSE

Two lines after the body of the letter, use one of the following complimentary closes:

Sincerely,
Sincerely yours,
Very truly yours,
Yours very truly,

Note that only the first word is capitalized and that all complimentary closes are followed by a comma.

SIGNATURE

Type your full name four lines below the complimentary close. Sign your name neatly in ink. Some organizations prefer that you add your position below your signature:

Very truly yours,

Chester Hall

Chester Hall
Manager, Procurement Department

REFERENCE INITIALS

If someone else types your letters, the reference initials identify both you and the typist. The initials appear a few lines below the typed signature. Generally, the writer's initials—which always come first—are capitalized, and the typist's initials are lowercased. For example, if Marjorie Conner wrote a letter that Alice Wagner typed, the reference initials would be MC/aw. See Figure 5-1 for an example.

NOTATIONS

An *attention line* is used if you cannot address a letter to a particular person:

Attention: Director, Biochemical Laboratories

The attention line is placed two lines below the inside address.

A *subject line* contains either a project number (*Subject: Project 65004*) or a brief phrase defining the subject of the letter (*Subject: Price Quotation for the R13 Submersible Pump*). The subject line is placed two lines below the attention line (or the inside address).

An *enclosure notation* is used if the envelope contains any documents other than the letter itself. For one enclosure, type:

Enclosure

For more than one enclosure, indicate the number of separate items (not the number of pages):

Enclosures (4)

The enclosure notation is placed a few lines below the reference line.

A *copy line* is used to indicate that you are sending carbon copies or photocopies of the letter (or sending it by electronic mail) to other people. The copy line is placed several lines below the enclosure notation:

cc: Robert Paul, Engineer Class 3

Some organizations do not include the recipient's position in the copy line.

Figure 5-1 shows the elements of a letter. The letter shown is presented

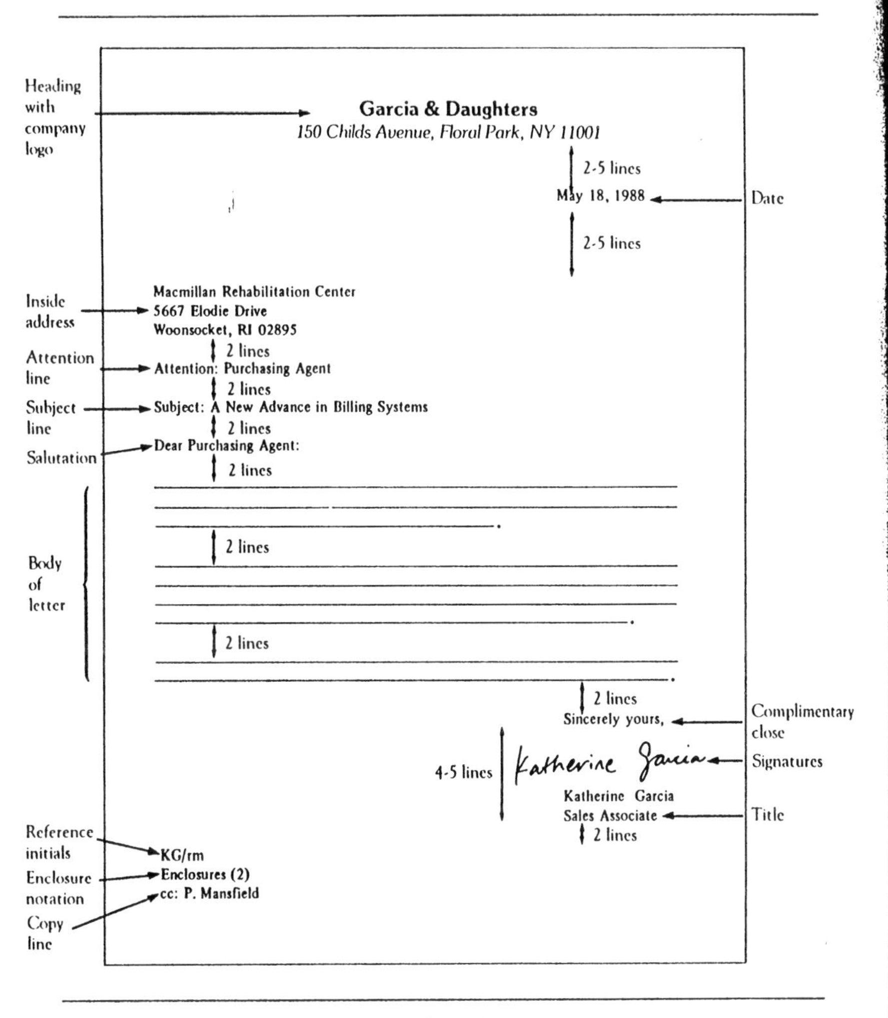

Figure 5-1. Sample Letter in Modified Block Format

in the *modified block format*, the most popular format. Two other formats should be mentioned:

1. *Modified block format with paragraph indentions.* In this format, the first line of each paragraph is indented four or five spaces.
2. *Full block format.* In this format, all the elements begin at the left margin.

Memos

Many memos are brief, consisting of one or two paragraphs. However, longer memos are now common because writers are increasingly using the memo format rather than the report format. Reports are often cumbersome and time-consuming to distribute because in many organizations company policy requires that they be signed by a number of officials.

The following discussion concentrates on longer memos.

Identifying Information

Almost all memos have five elements at the top: the logo or a brief letterhead of the organization and the "To," "From," "Subject," and "Date" headings. Some organizations have a "copies" or "cc" (carbon copy) heading as well.

If your organization permits it, indicate the job positions of both yourself and your readers. This will clarify things for the other readers as well as for anyone who wants to reconstruct the writing situation in the future. List the readers either alphabetically or in descending order of organizational rank.

When you create the subject heading, make it as clear as possible. Instead of writing "Tower Load Tests," write "Results of Tower Load Tests." Otherwise, your readers will not know whether the memo is about the date, the location, the methods, the results, or any number of other factors related to the tests.

As in a letter, the date should be written with the month expressed as a word, not as a number.

Following is an example of the identifying information section of a memo:

To: B. Pabst, Director of Operations
cc: William Weels, Comptroller
Victoria Chu, Personnel Department
From: J. Alonso, Chemical Engineering Department
Subject: Recommendations for Employee Stock Option Plan
Date: 17 August 1988

Second and subsequent pages of memos are typed on plain paper and include the following information at the top of the page: the name of the principal reader, the date, and the page number.

Memo to: B. Pabst 17 August 1988 page 2

Some organizations place the carbon-copy notation at the end of the memo. If a memo is addressed to a large number of readers, some organizations write "See Distribution" in the "To" space, and then include the distribution list at the end of the memo.

Purpose Statement

The purpose statement announces clearly the purpose of the memo. This statement gives the reader a clear idea of why you are writing and what you want him or her to do. In addition, writing the purpose statement forces you to clarify your purpose. Several examples of purpose statements follow.

I want to tell you about a problem we're having with the pressure on the main pump, because I think you might be able to help us.

The purpose of this memo is to request authorization to travel to the Brownsville plant Monday to meet with the other quality-control inspectors.

This memo presents the results of the internal audit of the Phoenix branch that you authorized March 13, 1988.

Summary

Along with the purpose statement, the summary forms the core of the memo. It helps all the readers to follow the subsequent discussion, enables executive readers to skip the rest of the memo, and serves as a convenient reminder of the main points. Following are some examples of summary statements:

The conference was of great value. The lectures on new coolant suggested techniques that might be useful in our Omega line, and I met three potential customers who have since written inquiry letters.

The analysis shows that lateral stress caused the failure. We are now trying to determine why the beam could not sustain a lateral stress weaker than that for which it was rated. See section 3b below for a discussion of our investigation methods.

As the second example shows, the summary can direct the readers' attention to portions of the full discussion. Although the summary precedes the full discussion, it is easier to write the full discussion first and then summarize it.

Detailed Discussion

Your subject will determine the nature of the detailed discussion. Your analysis of the audience might call for a background section followed by a problem-methods-solution structure. Or you might use a chronological, spatial, comparison-contrast, or any number of other patterns or combinations of patterns.

Remember that a memo is part of the official documentation of a project and therefore ought to be self-sufficient. Include enough background and discussion for future readers to understand it.

It is a good idea to use headings liberally throughout the discussion to help your readers.

Action Steps

Often, a memo requires some follow-up action, either by the writer or the readers. Creating an action section at the end of the memo helps everyone remember the tasks that have to be carried out. Be sure to indicate *who* is to do *what*, and *when* it is to be done by.

Following are two examples of action-step sections.

ACTION

I would appreciate it if you would work on the following tasks and have the results ready for the meeting on Monday, June 9.

1. **Henderson to recalculate the flow rate.**
2. **Smith to set up a meeting with the regional EPA representatives for some time during the week of February 13.**
3. **Falvey to ask Armitra in Houston for his advice.**

ACTION

To follow up these leads, I will do the following this week:

1. **Send the promotional package to the three companies.**
2. **Ask Customer Relations to work up a sample design to show to the three companies.**

Figure 5-2 incorporates the memo elements discussed in this section.

To: Sam Phillips, Director of Quality Assurance
From: Harry Chakowsky, Engineering Staff
Subject: Test Results on Ignitor Switch Alarm Control Cabinets 4A and 4B
Date: June 3, 1988

PURPOSE: to convey the results of the recent tests on the ignitor-switch alarm-control cabinets 4A and 4B, and to recommend remedial action.

SUMMARY: the cabinets passed all the tests except the electrical inspection: 1 of the 20 photohelic-gauge switches had an unexplainable slow response. I recommend that the switch be replaced so that the cabinet can be delivered to the client. See *Electrical Inspection*, below, for details of the test methods and results.

TEST METHODS AND RESULTS: Ignitor-switch alarm-control cabinets 4A and 4B were tested on June 1, 1988. Following is a discussion of the test methods and result.

Electrical Inspection
Result: On Cabinet 4B, Dwyer Photohelic Gauge Switch C-1 on Drawing E 6004 experienced a 7-second delay during the increasing cycle. The cause of this delay could not be determined by the vendor's Engineering Department in conjunction with our QA Inspectors. I *recommend* that, because Engineering decided this delay is unacceptable, we replace the switch so that we can deliver the cabinet to the client.

Method: I energized each of the switches to determine if the high and low switching was working properly.

Air Leakage Test
Result: All connections were approved.
Method: I tested each of the 20 instrument tubing circuits in each cabinet by:

1. disconnecting the fitting entering the photohelic gauge switch
2. attaching an air-pressure line to the fitting
3. capping the input connection on the top exterior level of the cabinet
4. opening the two Whitey brass shutoff hand valves in each circuit
5. applying 20 psig air pressure to each circuit
6. applying liquid LEK-TEC to each threaded joint, including the valve stems
7. retightening the joints that showed air leakage

Point-to-Point Continuity and Leakage-to-Ground Tests
Result: All wires and insulation were approved.
Method: I examined the following items for abnormalities:

1. identification tagging at each wire end
2. wire bundles
3. insulation on wires
4. placement of wires

If you have any questions about the inspection, please call me at X2168.

Figure 5-2. Sample Memo

EXERCISES

1. The following letter was written by a snowmobile dealer to a customer who brought his snowmobile in for servicing under the warranty. The letter is full of letter clichés, violates the "you attitude," and contains writing errors. Rewrite the letter so that it communicates more effectively.

Dear Mr. Smith:

Referring to your letter regarding the problem encountered with your new Eskimo Snowmobile. Our Customer Service Department has just tendered its report.

It is their conclusion that the malfunction is caused by water being present in the fuel line. It is our unalterable conclusion that you must have purchased some bad gasoline. We trust you are cognizant of the fact that while we guarantee our snowmobiles for a period of not less than one year against defects in manufacture and materials, responsibility cannot be assumed for inadequate care. We wish to advise, for the reason mentioned hereinabove, that we cannot grant your request to repair the snowmobile free of charge.

Permit me to say, however, that the writer would be pleased to see that the fuel line is flushed at cost, $30. Your Eskimo would then give you many years of trouble-free service.

Enclosed please find an authorization card. Should we receive it, we shall endeavor to perform the abovementioned repair and deliver your snowmobile forthwith.

2. R. Golen, an electrical engineer, was asked by the head of her department, K. Murphy, to recommend the best solution to the problem of inadequate pull strength in the LSI terminals that their company manufactures. Rewrite the following memo so that it will be easy for all the readers—including the managers listed in the "cc" notation—to understand.

To:	K. Murphy
From:	R. Golen
Subject:	Testing of Continuous Solder Strip Alternative for Large-Scale Integrated Terminals
cc:	J. A. Jones M. H. Miller

We ordered samples of continuous solder strips in three thicknesses for our testing: 1.5 mils, 2.5 mils, and 4.0 mils. Then, we manufactured each thickness into terminals to test for pull strength.

The 1.5-mil material had a test strength of 1.62 pounds, which is above our goal of 1.5 pounds. But 30 percent of these terminals did not meet the goal. The 2.5-mil material had an average pull strength of 2.4 pounds, with a minimum force of 1.65 pounds. The 4.0 mils material had an average pull

strength of 2.6 pounds per terminal, with a minimum of 1.9. Even though there was 60 percent more solder available than with the 2.4-mil material, the average pull strength increased by only 8 percent.

We concluded from this that the limit to the pull strength of the terminal is dependent on the geometry of the terminal, not on the amount of solder.

For this reason, we believe that the 2.5-mil material would be the most cost-effective solution to the problem of inadequate pull strength in our LSI terminals

Please let me know if you have any questions.

6

Elements of a Report

This chapter discusses some of the common elements found in technical reports. Most organizations have their own preferences about report elements. Therefore, you should check to see whether your organization has a style guide that describes them and gives examples. If it doesn't, study actual reports to see how the different elements are used.

Most completion reports contain many, if not most, of the following elements:

letter of transmittal
title page
abstract
table of contents
executive summary
body
references
appendixes

These elements are not written in the same order in which they are presented here. The letter of transmittal, for instance, is the first thing the principal reader sees, but it was probably the last item to be created. The reason is simple: the transmittal letter cannot be written until the document to which it is attached has been written. Many writers like to include in the first paragraph of the transmittal letter the title of the document; therefore, even the title has to be decided upon before the letter can be written. In the same way, the title cannot be decided until the body of the report is complete, because the title must reflect accurately the contents of the report.

The body of the report generally is written before any of the other items. After that, the sequence makes little difference. Many writers create the two summaries—the executive summary and the abstract—then the appendixes, and finally the table of contents, title page, and letter of transmittal.

Report Elements and the Word Processor

The task of assembling the formal elements of a report is much simpler if you can use a word processor. A word processor lets you see exactly how long each element is; therefore, you can easily expand it or contract it to meet any length requirements you might have.

In addition, the copy function on a word processor helps you create the different summaries and the transmittal letter. You can make a copy of the body of the report and then eliminate the details that do not belong. Creating the report elements by cutting material from the body is not only faster than writing them from scratch; it is also more accurate, for you don't introduce any technical errors.

Letter of Transmittal

The letter of transmittal introduces the purpose and content of the report to the principal reader. Generally written as a letter (although sometimes as a memo), it establishes a courteous tone for the report. Most transmittal letters contain the following five elements:

1. a statement of the title, and, if necessary, the purpose of the report
2. a statement of who commissioned the report, and when
3. a statement of the methods used in the project (if they are noteworthy) or of the principal results, conclusions, and recommendations
4. an acknowledgment of any assistance the writer received in preparing the materials
5. a gracious offer to assist in interpreting the materials or in carrying out further projects

Figure 6-1 is an example of an effective letter of transmittal.

Title Page

A typical title page consists of three elements:

1. a title that expresses the subject and purpose of the report, such as "Choosing a Microcomputer: A Recommendation for Allied Trucking" or "An Analysis of the Kelly 113 Packager." The title is centered and typed about one-third down the page. It is usually typed in uppercase letters without quotation marks or underlining.
2. the names of the principal reader and the writer, several lines below the title:
   Submitted to: Mr. Joseph Landry, President
   Garcia Associates

   Submitted by: William DeLord, Chief
   Operations Department
3. the date of submission, centered about two-thirds of the way down the page.

NYACK CONSTRUCTION
West Nyack, NY 11958

September 19, 1988

Mr. Wayne Endriss
Director of Operations
Nyack Utility Company
West Nyack, NY 11948

Dear Mr. Endriss:

Enclosed is the report on the site study of West Commack that you authorized on September 3, 1988.

Unfortunately, our conclusion is negative. An analysis of the site reveals an unsatisfactory soil composition and consistency for the [illegible] unavailable for our proposed facility. Given these two factors, I recommend that the site search be actively pursued.

I would be happy to answer any questions about the report, and I look forward to working on future site studies.

Yours very truly,

Carol Raff

Carol Raff, P.E.
Civil Engineering Department

Figure 6-1. Letter of Transmittal

Abstract

An abstract is a brief technical summary of a report—usually no more than 200 or 250 words. It is addressed primarily to readers who are familiar with the subject and who want to know whether to read the full report. Two kinds of abstracts are commonly written today: descriptive and informative.

A descriptive abstract describes the problem or opportunity that led to the project and then lists the topics covered in the report. It does not provide the major findings.

An informative abstract provides the findings—that is, the results, conclusions, and recommendations. Informative abstracts are generally more popular than descriptive abstracts. However, for documents such as descriptions of procedures, descriptive abstracts are necessary.

Figure 6-2 is an example of a descriptive abstract.

ABSTRACT

"Design of a Radio-Based System for Distribution Automation"
By Brian D. Crowe

At this time, power utilities' major techniques of monitoring their distribution systems are after-the-fact indicators, such as interruption reports, meter readings, and trouble alarms. These techniques are inadequate in two ways. One, the information fails to provide the utility with an accurate picture of the dynamics of the distribution system. Two, after-the-fact indicators are expensive. Real-time load monitoring and load management would offer the utility both system reliability and long-range cost savings. This report describes the design criteria we used to design the radio-based system for a pilot program of distribution automation. It then describes the hardware and software of the system.

Figure 6-2. Descriptive Abstract

This descriptive abstract could be turned into an informative abstract by replacing the last sentence with the findings, as shown in Figure 6-3.

The distinction between descriptive and informative abstracts is not absolute. Sometimes, you will combine elements of both in a single abstract. For instance, you are writing an informative abstract but the report includes 15 recommendations, which are far too many to list. You might decide to identify the major results and conclusions, as you would in any informative abstract, but state that the report contains numerous recommendations, as you would in a descriptive abstract.

ABSTRACT

"Design of a Radio-Based System for Distribution Automation"
by Brian D. Crowe

At this time, power utilities' major techniques of monitoring their distribution systems are after-the-fact indicators, such as interruption reports, meter readings, and trouble alarms. These techniques are inadequate in two ways. One, the information fails to provide the utility with an accurate picture of the dynamics of the distribution system. Two, after-the-fact indicators are expensive. Real-time load monitoring and load management would offer the utility both system reliability and long-range cost savings. This report describes the design criteria we used to design the radio-based system for a pilot program of distribution automation. The basic system, which uses packet-switching technology, consists of a base unit (built around a personal computer), a radio link, and a remote unit. The radio-based distribution-monitoring system described in the report is more accurate than the currently used after-the-fact indicators; it is small enough to replace the existing meters; it would pay for itself in 3.9 years; and it is simple to use. We recommend installing the basic system on a trial basis.

Figure 6-3. Informative Abstract

Table of Contents

Once you have created a clear system of headings within the report, transfer them to the contents page. Use the same format—capitalization, underlining, indention, and outline-style or decimal headings—that you use in the text.

If you are using a word processor, you have more format choices: boldface, italics, shadowed letters, different sizes of type, and so forth. Whether you are using a typewriter or a word processor, however, consider some basic aspects of formatting:

1. *Size*. Larger letters are more emphatic than smaller letters. Uppercase letters are more emphatic than lowercase. Boldface and italics are more emphatic than plain script. And underlining is more emphatic than not underlining. Using a standard typewriter, you can create up to six hierarchical levels using capitalization and underlining:

```
CATHODE RAY TUBES
CATHODE RAY TUBES
Cathode Ray Tubes
Cathode Ray Tubes
cathode ray tubes
cathode ray tubes
```

   Keep in mind, however, that most readers will become confused if you use more than four or five levels.

2. *Indention*. In general, more emphatic items begin closer to the left-hand margin. Subheadings, therefore, are indented four or five spaces.

```
First-level heading
    Second-level heading
        Third-level heading
```

   Many writers today use indention not only in the table of contents but in the text as well. For instance, the text that follows a second-level heading is indented the same five spaces as the heading itself.

```
First-level heading
..........................................................................
..........................................................................
      Second-level heading
      ....................................................................
      ....................................................................
```


   This indention reminds the reader of the level of the text.

   One exception to the principle of indention regards first-level headings. Some writers like to center first-level headings for emphasis:

```
              First-level heading
Second-level heading
    Third-level heading
```

3. *Outline-style and decimal-style headings*. The outline-style or decimal-style headings that you use in the text of a report should be transferred intact to the table of contents.

```
1.  FIRST-LEVEL HEADING
  A. Second-Level Heading
     1. Third-Level Heading
     2. Third-Level Heading
  B. Second-Level Heading

1.0 FIRST-LEVEL HEADING
    1.1 Second-Level Heading
        1.1.1 Third-Level Heading
        1.1.2 Third-Level Heading
  1.2 Second-Level Heading
```

A word processor makes it simple to transfer headings from the text of the report to the table of contents. Some software programs will actually do the job automatically. But even the simplest software helps you make the table of contents quickly and easily—and without introducing errors. Simply make a copy of the report. Then scroll through the copy, erasing the text. What is left are the headings. If they are inconsistent, you will note it immediately and thus be able to fix any problems both in the table of contents and in the text.

Once you have drafted the table of contents, check to see that the report contains enough headings and subheadings. For double-spaced type, the rule of thumb is that you have at least one heading or subheading on every page. Some pages might have two or three. The more headings and subheadings you have, the easier it will be for your readers to locate the information they seek. Make sure, however, that you do not violate a basic principle of logic by having only one subheading.

As you paginate the report, use lowercase Roman numerals (centered at the bottom of the page) for the pages before the table of contents. Use Arabic numbers (centered or on the right side at the top of the page) for everything after the table of contents. Do not number the table of contents page itself.

CONTENTS

Figure 6-4. Table of Contents

Figure 6-4 shows an effective table of contents. The report from which it is taken is titled "Investigation of the Feasibility of Replacing the Pneumatic Controller with an Electronic Proportional-Integral Controller on the Three-Tank Pressure System."

One other aspect of a table of contents should be mentioned here. If a report has many graphic aids, you might want to create a separate table of contents for them. If the list of illustrations can fit on the contents page, place it there; otherwise, create a separate page for it, and list it on the table of contents. Figure 6-5 shows a list of illustrations.

Executive Summary

An executive summary is a nontechnical condensation of a report addressed to a managerial and executive audience. Most managers and executives do not want or need a detailed understanding of the project. But they do need

LIST OF FIGURES AND TABLES

Figure 6-5. List of Figures and Tables

a basic understanding of the problem or opportunity that led to the project, as well as of the major findings. In many organizations, executive summaries are limited to one page, double-spaced. Other organizations use a two-page or 1000-word limit.

Keep in mind the following questions as you draft an executive summary:

1. *What was the problem or opportunity that motivated the project?* In describing problems you studied, focus on specific evidence. For most managers, the best evidence includes reduced costs or increased profits. Instead of writing, for example, that the equipment we are now using to cut metal foil is ineffective, write that the equipment jams on the average of once every 72 hours and that every time it jams we lose $400 in materials and $2000 in productivity as the workers have to stop the production line. Then, add up these figures for a monthly or annual total.

In describing opportunities you researched, use the same strategy. Let's suppose your company uses thermostats to control the heating and air conditioning. Research suggests that, if you had a computerized energy-management system, you could cut your energy costs by 20 to 25 percent. If your energy costs are $300,000 a year, you could save $60,000 to $75,000 annually. With these figures in mind, the readers have a good understanding of what motivated the study.

2. *What methods did you use to carry out the research?* In many cases, your principal reader does not care *how* you did what you did. He or she assumes you did it competently and professionally. However, if you think your reader is interested, include a brief description of your methods—it should be no more than a sentence or two.
3. *What were the main findings?* The findings are the results, conclusions, and recommendations. Sometimes, your readers understand your subject sufficiently and want to know your principal results—the data from your study. If so, provide them. Sometimes, however, your readers would not be able to understand the technical data or would not be interested. If that is the case, go directly to the conclusions—the inferences you draw from the data. For example, the results of the feasibility study on a computerized energy-management system might consist of a description of the different hardware and software that make up the system. If a brief description would be informative, include it. Otherwise, go directly to the conclusion: that is, answer the question, Would the system be cost effective?

   Finally, most managers want your recommendations—your suggestions for further action. Recommendations can range from the most conservative ("do nothing") to the most ambitious ("proceed with the project"). Often, a recommendation will call for further study of an expensive alternative.

After you have drafted the executive summary, give it to someone who has had nothing to do with the project. If that person finds it easy to read and understand, it will probably be effective for your readers.

Figure 6-6 is an example of an effective executive summary.

Note the differences between the executive summary and the informative abstract (Figure 6-3). The abstract focuses on the technical subject itself: whether the new radio-based system can effectively monitor the energy usage.

The executive summary concentrates on whether the system can improve operations *at this one company*. The executive summary describes in financial terms the symptoms of the problem at the writer's organization. After a one-sentence paragraph describing the system design—the results of

---

SUMMARY

At this time, we monitor our distribution system using after-the-fact indicators such as interruption reports, meter readings, and trouble alarms. This system is inadequate in two respects. First, it fails to provide us with an accurate picture of the dynamics of the distribution system. To ensure enough energy for our customers, we have to overproduce. Last year, we overproduced by 7 percent. This worked out to a loss of $273,000. Second, the system is expensive. Escalating labor costs for meter readers and the increased number of "difficult-to-access" residences have led to increased costs. Last year, we spent $960,000 reading the meters of the 12,000 "difficult-to-access" residences. This report describes a project to design a radio-based system for a pilot project on these 12,000 residences.

The basic system, which uses packet-switching technology, consists of a base unit (built around a personal computer), a radio link, and a remote unit.

The radio-based distribution monitoring system described in this report is feasible because it is small enough to replace the existing meters and because it is simple to use. It would provide a more accurate picture of our distribution system, and it would pay for itself in 3.9 years. We recommend installing the system on a trial basis. If the trial program proves successful, radio-based distribution-monitoring techniques will provide the best long-term solution to the current problems of inaccurate and expensive data collection.

Figure 6-6. Executive Summary

the study—the writer describes the findings in a paragraph. Note how the writer clarifies in the last paragraph how the pilot program relates to the overall problem described in the first paragraph.

Body

The body of a report typically contains an introduction, methods, and some or all of the following elements: results, conclusions, and recommendations. The kind and subject of the report will determine the exact content of these elements. Chapter 9, "Completion Reports," describes the body in detail.

References

Every organization uses a different format for citing references. If your organization has a style guide or uses an external style guide, refer to it. If not, refer to other reports for guidelines.

Appendix B, page 191, describes how to document sources.

Appendixes

An appendix is an item (except an index) that is attached to the end of a document. Appendixes are a convenient way to include material that is too bulky to be included in text or that might be of interest to only some of your readers—for example, maps, large technical diagrams, glossaries (lists of definitions of technical terms), computations, computer printouts, supporting documents, and so forth.

Appendixes, which are usually identified by letter (for example, *Appendix B*), are listed in the table of contents. Remember that an item in an appendix is designated "Appendix," not "Figure" or "Table," even if it would have been so designated had it appeared in the body of the report.

EXERCISES

1. The following excerpt is taken from the completion report of the study that is proposed in the Exercises at the end of Chapter 7. On the basis of this excerpt, write

   a. a transmittal letter from Edwin Korody, Food Scientist, Department of Food Development, to Dr. Davis Figgins, General Manager of that department
   b. a title page
   c. an informative abstract (for the body sections)
   d. an executive summary

II. DETERMINATION OF AN OPTIMAL CASEIN AND SOY PROTEIN MIXTURE

A. INTRODUCTION

The purpose of this research was to determine an optimal ratio of casein and soy protein for our product RatFeed. The present formulation contains a 10-percent weight of a complete protein source, casein. This formula has been used since the feed was designed in 1966. By December of 1986, a 25-percent price increase of casein is expected. This will directly affect the production cost of the product. In order to avoid a price increase, an alternative formulation was considered.

The nutritional quality of a protein depends on the quantity, availability, and proportions of the essential amino acids that make up the protein. A complete protein, such as casein, can provide growth and a positive nitrogen balance because it contains all the essential amino acids for a growing animal. The Amino Acid Score or Amino Acid Index method (Oser 1959) shows that casein provides a ratio of essential amino acids similar to that required by a growing rat. On the other hand, soy protein appears limited in the sulfur-containing amino acids: L-cisteine and methionine. Soy is an incomplete protein, unable to promote growth by itself, and it is of inferior protein quality when compared to casein. The Amino Acid Index also shows that adding casein to a soy-protein diet should complement the deficiency of soy's limiting amino acids, increasing its protein quality. But, the Amino Acid Index method makes an unreliable prediction, since it does not account for the absorption, availability, and metabolic interactions of the amino acids.

Bioassays, on the other hand, compensate for these factors by measuring the efficiency of the biological use of dietary proteins as sources of the essential amino acids under standard conditions. Many of these biological methods are based on the effects of the quality and quantity of protein on growth performance of growing animals. Several biological assays have been proposed to measure protein quality. Protein Efficiency Ratio was chosen for this research because it is widely used, is the official method in the United States and in Canada, and is simple and inexpensive.

B. METHODOLOGY

The protein efficiency ratio (PER) method was first used in 1917 by Osborne and Mendel in their studies to establish protein quality (Osborne et al. 1919). This method requires that energy intake be adequate and the protein be fed at an adequate but not excessive level in order to promote growth. This is compatible with both the 10-percent protein level and the caloric content of RatFeed. PER is a measure of weight gain of a growing animal divided by its protein consumption. The official AOAC procedure requires a number of factors to be modified in order to standardize the reference casein and test diets. These factors include the level of the protein intake; type and amount of dietary fat; fiber levels; strain, age, and sex of rats; assay lifespan; room conditions, etc. All the biological factors were followed in our research, but instead of using the official casein diet our current RatFeed was used as our standard.

1. *Diet Preparation*

Before the arrival of the rats, 4 kg of each diet were prepared, with the following protein contents:

Diet 1: 25% casein, 75% soy protein
Diet 2: 50% casein, 50% soy protein
Diet 3: 75% casein, 25% soy protein
Diet 4: 100% casein (our current feed)

With a fixed total protein level of 10 percent, the diets were isocaloric. All other factors and ingredients of the diets were fixed at the current proportions of our RatFeed product. The following is the standard percentage weight formulation used for the reference product:

INGREDIENTS	PERCENTAGE WEIGHT
Protein (100% casein)	10
Starch	72
Oil (corn-cottonseed)	9
Mineral mixture[a]	5
Water	2
Cellulose (Pax-Cell)	1
Vitamin mixture (McNeil Lab.)[a]	1
	100%

[a] The composition of the salt and the vitamin mixture are given in Appendix A.

2. *Rat Bioassay*

We used 20 weanling male, Sprague-Dawly rats, weighing 48 to 60 grams. The animals were housed in individual galvanized steel cages (7″ wide, 7″ high, 15″ long) in Room 505 of the Animal Research Department. In Room 505, temperature (72°–75°F) and lighting (12 hrs light/12 hrs dark) were controlled. Assay groups were assembled in lots of 5 rats per diet in a random manner so that the weight difference was minimized. Throughout a period of 28 days, feed and water were provided ad libitum, after a fasted period of 36 hours.

Food consumption records and weight gains were recorded for individual rats every seven days. Data were collected in an automatic electronic balance. Food disappearing, considered as food intake, was determined as the difference beween the weights of the food cup before and after filling, with corrections made for food spilled. Protein intake was calculated by multiplying the percentage of protein in the diet (10%) by the total food intake, divided by 100. PER was calculated as the weight gain divided by

the amount of protein consumed. The raw data for PER, weight gain, food intake, and protein intake are given in Appendix B.

C. RESULTS

The PER, weight gain, food intake, and protein intake values for the assay are shown in Table 1. Data in Table 1 were statistically manipulated

Table 1

Weight Gain, Food Intake, Protein Intake, and PER for RatFeed and Test Diets

Diet	*Weight Gain (grs)*	*Food Intake (grs)*	*Protein Intake (grs)*	*PER*
25% casien-75% soy prot.	52.96 ± 7.62	331 ± 70.6	33.1 ± 7.06	1.6 ± 0.22
50% casein-50% soy prot.	88.80 ± 5.56	370 ± 50.2	37.0 ± 5.25	2.4 ± 0.38
75% casein-25% soy prot.	101.79 ± 9.84	352 ± 43.4	35.1 ± 4.34	2.9 ± 0.21
100% casein (RatFeed)	93.86 ± 5.91	362 ± 32.3	36.1 ± 3.2	2.6 ± 0.45

* Mean ± Standard Deviation

using the MINITAB system. A two-sample t-test was used to check for significant differences of the means (χ = 0.05). PER values range from 1.6 to 2.6. The PER values from the 50-percent casein, 50-percent soy diet and our standard feed proved by this statistical manipulation to have no significantly different means. The other two diets have different PER values from our standard. The food intake, and subsequently the protein intake, did not vary in the standard nor in the test diets.

D. CONCLUSION

Our results show that a mixture of 50-percent casein and 50-percent soy protein has the same PER value (around 2.5) as our standard product. Therefore, it can be concluded that the two have the same biological protein quality. Moreover, this optimal protein ratio could substitute for casein in RatFeed without altering the expected nutritional performance of the product. At the same time—as described in the executive summary—

modifying our current protein to 50-percent casein, 50-percent soy will reduce by 27 percent, or $3,200 per year, our current production costs for this product.

E. RECOMMENDATION

If the Production and Quality Control managers have no objection, the change in formulation of our current feed might be a feasible solution to avoid the increase in its production cost.

REFERENCE

Crowe, B.D. 1985. Design of a radio-based system for distribution automation. Unpublished document.

7

Proposals

A proposal is an offer to provide a product or service in exchange for money or some other asset. For example, when a police department wishes to purchase a fleet of cars, the automobile manufacturers interested in winning the contract submit proposals stating the cost, specifications, and delivery dates of their products. The police department then grants the contract to the manufacturer who best meets its needs. This kind of proposal is called an *external proposal*, because the supplier is an outside organization.

An *internal proposal* is submitted to someone within the writer's own organization. For internal proposals, the payment usually takes the form of time rather than money. For instance, when an employee proposes that he or she be permitted to carry out a study, the employee asks for time and related resources—such as computer time, equipment, and secretarial help—to do the work. Some internal proposals simply request authorization to make a purchase or implement some change in company operations. Internal proposals range in scope from the informal—memos or even phone calls—to the formal. Regardless of form and length, however, internal proposals must be as carefully thought out and presented as external proposals.

The Role of the Proposal

A vast network of contracts spans the working world. The United States government, the world's biggest customer, spent about $250 billion in 1986 on work contracted to organizations that submitted proposals. The defense and aerospace industries, for example, depend almost totally on government contracts. But proposal writing is by no means limited to government contractors. One auto manufacturer buys engines from another, and a company that makes spark plugs buys its steel from another company. Most services, too, are contracted out. Organizations retain other organizations to provide computer services, security, maintenance, training, and dozens of other services.

When an organization wants to purchase a product or service, it issues one of two kinds of documents. An IFB, an information for bid, is a request for bids for a standard item. If the government, for instance, wishes to purchase one million pencils, it issues a simple IFB. The company that offers to provide the product for the best price wins the contract.

An RFP, a request for a proposal, is issued for a customized product or service. The police department that wishes to buy a fleet of cars wants different engines, cooling systems, suspensions, and upholstery than are

offered on the standard consumer automobile. The RFP specifies what the police department wants. The supplier that, at a reasonable price, can provide the automobiles that most closely resemble the specifications wins the contract. Sometimes, the RFP is a more general statement of goals. The purchaser in effect asks the suppliers to create their own designs or describe how they will achieve specified goals. The supplier that offers the most persuasive proposal will probably be awarded the contract.

IFBs and RFPs are published in newspapers, in magazines, and in the government's publication *Commerce Business Daily*.

In most cases, the world of contracts is a buyer's market: there are many organizations that want to provide the products and services. Therefore, proposal writers must show that they understand the readers' needs and are willing and able to keep their own promises.

Understanding your readers' needs is critical. Although this point might seem obvious, people who evaluate proposals, including government officials, private foundation officers, and managers in small corporations, agree that most proposals fail to meet the readers' needs. Sometimes, the RFP was unclear or the proposal writer misunderstood it. Sometimes, the writers knew they couldn't meet the readers' needs but hoped that the readers wouldn't notice or that no other supplier came closer. But a proposal that doesn't solve the specified problem will probably be tossed out quickly.

Therefore, determine as precisely as possible your readers' needs. Study the RFP or other description. If you don't understand something in it, ask the issuing organization. They will be happy to help, because an ineffective proposal wastes everyone's time. If you are writing an internal proposal, get all the information you can on how the current situation is reducing productivity or decreasing the quality of your organization's product or service. If you are writing about an opportunity, get all the facts on how this opportunity would increase quality or reduce costs.

Once you have shown your readers that you understand the problem or opportunity, describe in detail what you plan to do about it. Specify what methods and equipment you would use. Create a complete picture of how you would get from the first day of the project to the last. Don't expect your readers to trust you. They want to see details.

Don't overlook the question of professionalism. Although a lot of organizations and individuals have the expertise to carry out a project, some of them lack the pride, ingenuity, and perseverance to get the job done. Explain how you plan to monitor your progress and guarantee your work. Describe similar projects you have completed successfully.

Elements of a Proposal

If your readers have provided an RFP or some other set of guidelines, follow it to the letter. If no guidelines exist, use the following structure:

1. summary
2. introduction
3. proposed program
4. qualifications and experience
5. budget
6. appendixes

As is usually the case with technical documents, the sequence of composition is not the same as the sequence of presentation. The first section to write is the proposed program. Until you know what you propose doing, you cannot write the summary or the introduction. After you have drafted the proposed program, proceed to the end: the qualifications and experience, budget, and appendixes. Once you have completed all these items, introduce the proposal and draft the summary.

Summary

For any proposal of more than a few pages, provide a summary. The summary is crucial because in many cases it will be the only item the readers study in their initial review. Some organizations impose a length limitation, such as 250 words.

To write an effective summary, reduce the problem/opportunity statement from the introduction to two or three sentences. Then devote several sentences to your proposed program, and one or two to your qualifications and experience. Finally, state your completion date and the total budget.

Figure 7-1 shows an effective summary taken from a proposal submitted by a metal foundry to a college to do some modifications to a piece of equipment, an air gun, owned by the college.

Introduction

The body of the proposal begins with an introduction, which defines the background and the problem or opportunity.

The background is the context: that is, the relationships or events that led to the discovery of the problem or opportunity. In discussing the

---

Summary

American Metal Foundry, Inc., proposes to modify St. Thomas College's pneumatic high-pressure air gun (PHPAG) to increase its usefulness in destructive-testing research. The project entails redesigning the barrel of the gun and designing and manufacturing the receiver tank, which would hold the test specimens and catch the flying projectiles.

These modifications would enable St. Thomas College to carry out sophisticated destructive-testing research in its materials laboratory.

This project, which would take 17 days, would cost $8066. The unit would be guaranteed to meet specifications after it is delivered and installed.

Figure 7-1. Summary Section

problem or opportunity itself, use monetary terms. Your proposal itself will include a budget; you want to be able to convince your readers that spending money on what you propose is a wise investment. Don't say that a design problem is slowing down production; say that it is costing $4500 a day in lost productivity.

Figure 7-2 is the introduction to the air-gun proposal.

Proposed Program

In describing what you plan to do, be specific. Don't write that you plan to "gather data and analyze it." How will you gather it? How will you analyze it? Every word that you write—or don't write—will give your readers evidence on which to base their final decision.

If your project concerns a subject that has been written about in the professional literature, show your readers that you are familiar with the scholarship by referring here (or earlier, in the introduction) to the pertinent studies. For instance:

> **Carruthers (1987), Harding (1987), and Vega (1988) have demonstrated the relationship between acid-rain levels and groundwater contamination. None of these studies, however, included an anal-**

Introduction

American Metal Foundry, Inc., proposes to modify St. Thomas College's pneumatic high-pressure air gun (PHPAG) to increase its usefulness in destructive-testing research. The project entails redesigning the barrel of the gun and designing and manufacturing the receiver tank, which would hold the test specimens and catch the flying projectiles.

St. Thomas has a 15-ft PHPAG capable of firing projectiles at speeds in excess of 1,200 ft/sec. A specimen of a desired material is placed in front of the barrel. A projectile is then loaded and fired at the specimen. The effect of the impact is then measured. The results of this testing yield valuable information about the material.

The problems that this proposal addresses are as follows:

1. Although the barrel of the PHPAG has an inside diameter large enough to accommodate a 2.5-in.-diameter projectile, the barrel has an insert that is used to fire 0.60-caliber shells. This insert prevents the use of larger shells.
2. The PHPAG does not have a receiving tank. Therefore, it cannot accommodate the larger projectiles necessary for more advanced destructive testing.

Modifying the barrel of the PHPAG and building a receiving tank will allow St. Thomas to carry out high-level destructive testing in its materials laboratory.

Figure 7-2. Introduction Section

ysis of the long-term contamination of the aquifer. The current study will consist of . . .

You might include just one reference to recent literature. However, if your topic is the subject of substantial research, you might devote several paragraphs or even several pages to a discussion of recent scholarship. Figure 7-3 shows the proposed program for the air-gun project. In this example, the discussion of the proposed project is a general description of the process that the writer's organization will perform. Note that this proposed procedure does not discuss any research, for the project does not involve any original research.

Proposed Procedure

The project entails two major stages:

1. redesigning the barrel of the gun
2. designing and manufacturing the receiver tank, which would hold the test specimens and catch the flying projectiles.

1. Redesigning the Gun

The barrel of the gun has an inside diameter large enough to accommodate a 2.5-in.-diameter projectile. However, the barrel has been fitted with an insert used to fire 0.60-caliber shells. This insert prevents the use of larger shells. We propose to remove this insert and re-machine the barrel so that it can accommodate the larger projectiles.

2. Designing and Manufacturing the Receiver Tank

Adding a receiver tank involves several steps. We propose to design and manufacture a tank large enough to tolerate the enormous impact involved in destructive testing. We would install the gun in one side of the tank. At the insertion point, we would install gaskets and flanges to keep the gun and tank airtight during firing. Finally, after testing the unit we would mount the tank to the floor in your laboratory. We would then repeat the testing to ensure that the unit performs according to specifications.

Figure 7-3. Proposed-Program Section

Qualifications and Experience

The more elaborate the proposal, the more substantial your discussion of qualifications and experience has to be. For a small project, a few paragraphs describing your technical credentials and those of your coworkers will suffice. For larger projects, the résumés of the project leader and the other important participants should be added.

External proposals should also include a discussion of the qualifications of the supplier's organization, focusing on similar projects completed successfully, technical support staff, equipment and facilities, management structure, and so forth.

Figure 7-4 is the qualifications-and-experience section of the air-gun proposal.

Qualifications

American Metal Foundry, Inc., has been performing high-quality metal-machining work since 1956. Our engineering staff consists of experts in stress analysis, mechanics, and machine design and fabrication. Our technicians combine state-of-the-art technical expertise and old-fashioned pride in their workmanship. In fact, we have won seven Certificates of Merit from the Association of Metal Foundries for our products.

Our team leader, Dr. Karen Mair, has over 14 years' experience in designing and fabricating metals and other materials. She is supported by a fully professional support staff of engineers, technicians, and clerical personnel. Dr. Mair has successfully completed projects for dozens of organizations, including Valley View Hospital, Parker State University, Hawkins Ford-Mercury, and Demling Nursery. For a complete listing of her credentials, see her résumé in Appendix B. Please see Appendix D for a full description of similar projects we have undertaken.

Figure 7-4. Qualifications-and-Experience Section

Budget

Good ideas aren't good unless they're affordable. The budget section of a proposal specifies how much the proposed program will cost.

A budget for a simple internal proposal can be a single sentence: "This study will take me about two days, at a cost of about $400, including secretarial time," or "The variable-speed recorder currently costs $225, with a 10-percent discount on orders of five or more."

For a more complicated internal proposal or for any external proposal, the budget is usually divided into two parts: direct costs and indirect costs. Direct costs include such expenses as salaries and fringe benefits, travel expenses, equipment, materials, and supplies. Indirect costs cover the intangible expenses called *overhead*: general secretarial expenses, utilities, maintenance, and so forth. Indirect costs are generally expressed as a percentage—ranging from less than 20 percent to more than 100 percent—of the direct costs.

Figure 7-5 is the budget from the air-gun proposal.

Budget Itemization

for period August 1, 1988 to August 17, 1988

Direct Costs

1. Salaries and Wages

Personnel	*Title*	*Time*	*Amount*
Karen Mair, Ph.D.	Design Engineer	2 weeks	$1900
Ed Smith	Chief Machinist	2 weeks	1200
	Typist	2 days	175
Total Salaries and Wages			$3275

2. Supplies and Materials

Tank	$2300
Miscellaneous Materials	400
Total Supplies and Materials	$2700
Subtotal	$5975

Indirect Costs

35% of $5975	$2091
Total Cost	$8066

Figure 7-5. Budget Section

Appendixes

Many different kinds of appendixes might accompany a proposal: descriptions of related projects, testimonials to the writers' skill and integrity, task schedules, descriptions of evaluation techniques, and the like. Often, the RFP specifies which of these items is required. Even if they are not required, however, many organizations like to provide them.

A task schedule is generally presented as a horizontal bar chart, called a *Gantt chart*. Figure 7-6 is the task schedule for the air-gun proposal.

Evaluation techniques vary widely. Some proposals call for evaluation by an outside agency, such as a testing laboratory or a university. Others call for evaluations, such as cost-benefit analyses, performed by the supplier

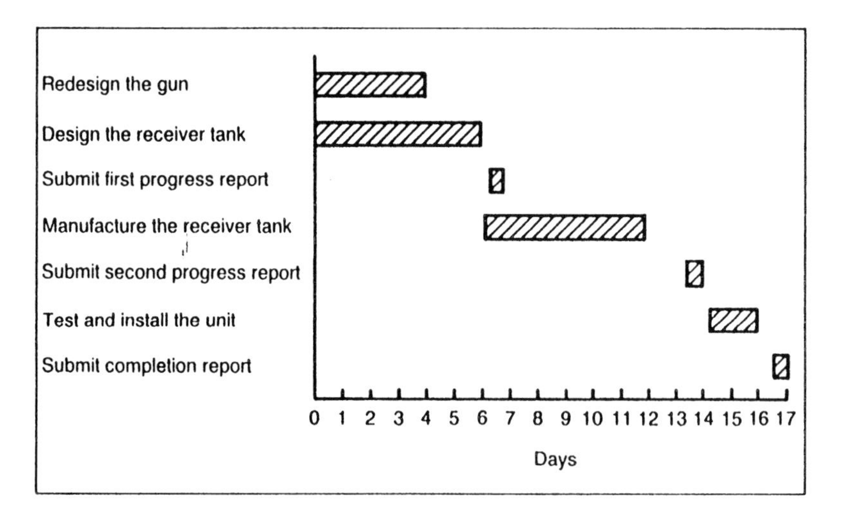

Figure 7-6. Task Schedule for a Proposal

itself. Evaluations can be *qualitative*—focusing on the improved quality of the product or service—or *quantitative*—focusing on improved productivity, for example. Evaluations can be *formative*—carried out while the project is still under way—or *summative*—carried out after the project is completed.

Because evaluation is such a complex subject, be sure you understand what the issuing organization has in mind before you submit a proposal.

Figure 7-7 is the evaluation-techniques section of the air-gun proposal.

Appendix A. Evaluation

We will submit two progress reports to St. Thomas College. We will submit one at the end of week one and one at the end of week two. In addition, we would welcome your inspection of the progress on the project at any time. We will guarantee that the unit will work according to the agreed specifications for at least one year once it is installed at your facility.

Figure 7-7. Evaluation-Techniques Section

EXERCISES

1. Write a proposal for a research project. Start by defining a technical subject that interests you. Using abstract journals and other bibliographic tools, create a bibliography of articles on the subject. Then make up a reasonable real-world context: for example, you could pretend to be a young civil engineer whose company is considering the purchase of a new kind of earth-moving equipment. Address the proposal to your supervisor, requesting authorization to investigate the advantages and disadvantages of this new piece of equipment.

2. The following internal proposal was written by a student nutritionist working for an animal-feed manufacturer.* In an essay, evaluate the effectiveness of the proposal, commenting on both the content and the expression.

October 25, 1985

To: Davis Figgins, General Manager

From: Edwin Korody, Food Scientist, Department of Food Development

Subject: Proposal to Determine the Best Casein and Soy Protein Mixture for the Product RatFeed

Problem

The expected 25-percent increase in the price of casein next month will directly affect our production costs because casein makes up 10 percent of the total weight of RatFeed. I would like to investigate an alternative formulation.

Introduction

Recent studies in nutrition suggest that a combination of high- and low-quality protein can be substituted for high-quality dietary proteins, such as casein, without hurting the overall protein quality (Pike and Brown 1984).

This memo proposes a study substituting a soy-casein mix for the casein portion of RatFeed. Soy is currently half the price of casein.

Protein quality will be measured using the protein efficiency ratio (PER)

* E. Korody. 1985. "Proposal to Determine the Best Casein and Soy Protein Mixture for the Product RatFeed." Unpublished document.

method. PER is a measure of weight gain of a growing animal divided by protein intake. The PER method was first used in 1917 by Osborne and Mendel in their studies to establish protein quality (Osborne et al., 1919). The PER bioassay method will be employed in the study because it is widely used, it is the official method used in the United States and Canada, and it is simple and inexpensive.

Proposed Procedure

The bioassay procedure outlined in the official FDA method (AOAC 1975) for conducting the biological evaluation of protein quality will be followed. Diets with the following protein compositions will be prepared prior to the arrival of the rats.

Diet 1: 25% Casein, 75% Soy
Diet 2: 50% Casein, 50% Soy
Diet 3: 75% Casein, 25% Soy
Diet 4: 100% Casein (RatFeed)

With a fixed protein level of 10 percent, the diets will be isocaloric. All other factors and ingredients of the diets will be fixed at the current proportions of our RatFeed product.

Four-kg. batches of each diet will be prepared and kept refrigerated to avoid spoilage or vitamin losses. Twenty weanling male, Sprague-Dawly rats, weighing 45 to 55 grams, 21 to 25 days of age, will be obtained from our commercial supplier (Crosby Laboratories, Media, PA 19220). The

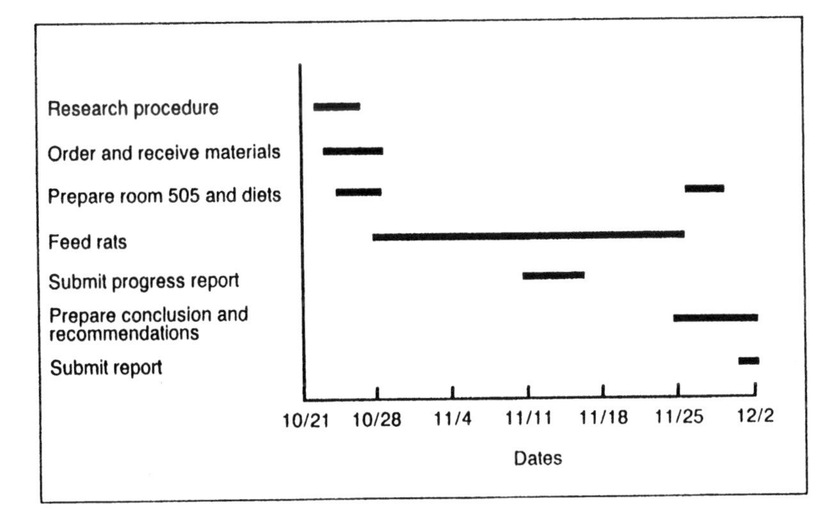

Task Schedule

animals will be housed in the individual galvanized steel cages (7″ wide, 7″ high, 15″ long) in Room 505 of the Animal Research Laboratory. In Room 505, temperature (72°–75°F) and lighting (12 hrs light/12 hrs dark) is easily controlled. Assay groups will be assembled in lots of five rats per diet in a random manner so that the weight differential is minimized. Throughout the 28-day test period, feed and water will be provided ad libitum, after a fasted period of 36 hours. Individual rat food consumption records and weight gains will be determined on a seven-day basis. Data will be collected in the automatic electronic balance. Upon conclusion of the 28-day feeding study, individual rat protein efficiency ratios (PER) will be calculated on the basis of individual weight gain and consumption of diets. Comparisons will be made between PER of the mixed-protein diets and PER of the reference diet to determine the optimal casein/soy ratio.

Budget

Direct Cost	Edwin Korody	35 hours at $5.00/hr.:	$175.00
	Typist	5 hours at $3.50/hr.:	17.50
	Rats		30.00
	Diets		25.00
		Subtotal	$247.50
Indirect Cost			
	20% of $247.50		$ 49.50
		Total cost	$297.00

Credentials

In my two years' working experience at Agricultural Foods Co., I have participated in several reformulations projects to meet the new FDA standards. During my educational training at Forster University, I took courses requiring animal bioassays. In addition, I spent six months of work-study in charge of the rat laboratory of Forster Nutrition graduate students.

References

AOAC. 1975. *Official Methods of Analysis* (12th Ed. 48.183) Association of Official Agricultural Chemist, Washington, D.C.

FDA. 1973 Federal Register 38:6951.

Osborne, T., L. Mendel and Edna Ferry. 1919. A method of expressing numerically the growth-promoting value of proteins. *J. Biol. Chem.* 37:223.

Pike, R. and M. Brown. 1984. *Nutrition. An Integrated Approach.* 3rd. Ed. John Wiley & Sons. New York. pp. 30–45, 736–742.

8

Progress Reports

A *progress report* communicates to a supervisor or a sponsor the current status of a project that has been begun but not yet completed. As its name suggests, a progress report is an intermediate communication submitted after the proposal (see Chapter 7) and before the completion report (see Chapter 9). The purpose of a progress report is to let the writer "check in" with the reader.

The Role of the Progress Report

A progress report usually describes a single project. A periodic report on the entire activities of a department or division of an organization is commonly called a *status report*.

A progress report is submitted according to a schedule: for big projects, every month or every quarter; for small projects, perhaps every week. The format varies, too. For big projects, it might be a formal report; for small projects, it might be a memo or even a phone call.

The progress report is the proposal updated in light of recent experience. On rare occasions, you might be able to report an absence of difficulties. But if unanticipated events have occurred and the expected result, the cost, or the schedule has to be revised, you have to explain clearly and fully what happened and how it will affect the overall project. Your tone should be objective, neither defensive nor casual. Unless ineptitude or negligence caused the problem, you are not to blame.

Regardless of what news you are delivering—good, bad, or mixed—your job is to provide a clear and complete record of your team's activities and to forecast the next stage of the project.

Elements of a Progress Report

Progress reports vary widely in structure, because of differences in format and length. Written as a one-page letter, a progress report is likely to be a series of traditional paragraphs. As a brief memo, it might also contain section headings. As a report of more than a few pages, it might contain the elements of a formal report.

Despite these differences, however, most progress reports share a basic structural pattern. The writer:

1. introduces the readers to the progress report by explaining the objectives of the project and providing an overview of the whole project

2. summarizes the progress report
3. discusses the work already accomplished and the work that remains to be done, speculating on the future promise and problems of the project
4. concludes by evaluating the progress of the project

Relevant appendixes are attached to the report.

When you write the progress report, leave the summary for last; you can't summarize what you haven't yet written. However, the other elements of the standard progress report can be written in the sequence in which they will be presented.

If you are using a word processor, start by making a copy of the proposal. Many of the elements of the progress report are taken directly from the proposal, or based on it. The problem or opportunity that led to the project, for instance, can often be presented intact. The discussion of past work is often a restatement of a portion of the proposed program, with the future tense changed to past. You might want to reproduce your task schedule—modified to show your current status—and your bibliography.

INTRODUCTION

This is the first monthly progress report on the project to develop a new testing device for our computer system. The goal of the project is to create a device that can debug our current and future systems quickly and accurately.

The project is devided into four phases:

I. define the desired capabilites of the device
II. design the device
III. manufacture the device
IV. test the device

Figure 8-1. Introduction to a Progress Report

Introduction

The introduction provides background information that orients the reader to the progress report. It identifies the document as a progress report, describes the problem or opportunity that led to the project, states the objectives of the project, and outlines the phases of the project.

Figure 8-1 is an example of an introduction to a progress report.

Summary

A progress report of more than a few pages often contains a summary, which briefly describes the accomplishments of the current reporting period and then comments on the current work. Note how the summary in Figure 8-2 calls the reader's attention to a problem described in greater detail later in the progress report.

SUMMARY

In November, Phase I of the project--to define the desired capabilities of the testing device--was completed.

The device should have two basic characteristics: versatility and simplicity of operation.

1. Versatility. The device should be able to test different kinds of logic modules. To do this, it must be able to "address" each module and ask it to perform the desired task.
2. Simplicity of Operation. The device should be able to display the response from the module being tested so that the operator can tell easily whether the module is operating properly.

Phase II of the project--to design the device--is now underway.

We would like to meet with Research and Development to discuss a problem we are having in designing the device so that it can "address" each module. Please see "Discussion" and "Conclusion" below.

Figure 8-2. Summary Section

Discussion

The discussion elaborates on the points mentioned in the summary. Of the several methods of structuring the discussion, the simplest is the past work–future work scheme. After describing the problem or opportunity that motivated the project, you describe the work that has been completed in the present reporting period and then sketch in the work that remains. This structure is easy to read and to write.

A second structure for the discussion is based on the tasks involved in the project. If the project requires that you work on several of the tasks at the same time, this structure is particularly effective, for it enables you to describe, in order, what you have accomplished on each task. Often, the task-oriented structure incorporates the past work–future work structure:

III. Discussion
- A. The Problem
- B. Task I
  1. Past work
  2. Future work
- C. Task II
  1. Past work
  2. Future work

Figure 8-3 shows the standard chronological progression, from the problem to past, present, and future work. Note that the writer uses a combination of generic and specific phrases in the headings.

Conclusion

Your reader will receive at least one more report: another progress report or a completion report. The conclusion of a progress report is therefore more transitional than final.

Your task in the conclusion is to evaluate your progress. In the broadest sense, you have one of two messages: things are going well, or things are not going as well as anticipated.

If things are going well, convey your optimism, but don't get carried away. Don't promise to complete the project early or under budget, for instance; new problems might arise that cost you time or money.

On the other hand, don't panic if the current situation isn't what you had hoped it would be at this point. Explain the situation clearly and

DISCUSSION

The Problem

Isolating and eliminating design and production errors has always been a top priority of our company. Recently, the sophistication of new computer systems we are producing has outpaced our testing capabilities. We have been asked to develop a testing device that can quickly and accurately debug our current and projected systems.

Currently, we have no technique for testing the individual logic modules of our new systems. Consequently, we cannot test the system until all of the modules are in place. When we do discover a problem, such as a timing error in one of the signals, we have to disassemble the system and analyze each of the modules through which the signal flows. Even after we have discovered a problem, we cannot know if that is the only problem until we reassemble the system and test it again. Although this testing method is effective--we are well within acceptable quality standards--it is very inefficient.

The solution to this problem is to develop a device for testing the logic modules individually before they are installed in the system. That is the overall objective of the project. Phase I of the project involved our determining the desired capabilities of this device.

Work Completed: Determining the Desired Characteristics of the Testing Device

The first characteristic of the testing device must be versatility. Over the next two years, we will be introducing three new systems. This rate of introduction is expected to continue at least through the 1980s. Although we cannot foresee the specific components of these new systems, of course, all are expected to incorporate the latest large-scale integration techniques, in conjunction with microprocessor control units. This basic structure will enable us to employ separate logic modules, each of which performs a specific function. The modules will be built on separate logic cards that can be tested and replaced easily. To achieve this versatility, the testing device should be able to "address" each module automatically and ask it to perform the desired task. Because the testing device will be asked to handle a large number of logic modules, it should be able to distinguish between the different modules in order to ask them to perform their appropriate functions.

The second characteristic of the testing device should be simplicity of operation: the ability to display for the operator the response from the module being tested. After the testing device transmits different command and data lines to the module, it should be able to receive a status or response word and communicate it to the operator.

Future Work

We are now at work on Phase II--designing the device to reflect these desired characteristics.

We are analyzing ways to enable the device to automatically "address" the various logic modules it will have to test. The most promising approach appears to be to equip each logic module with a uniform integrated circuit--such as the 45K58, a four-bit magnitude comparitor--that can be wired to produce a unique word that indicates the board address of that module.

The display capability appears to be a simpler problem. Once the module being tested has executed a command, it will generate a status word. The testing device will receive this status word by sending an enable signal to the status enable pin on the unit holding the module. Standard LCD indicators on the front panel of the testing device will display the status word to the operator.

Figure 8-3. Discussion Section

objectively. Don't withhold bad news, hoping that you'll figure out a way to fix things later. If your news is not good, at least give your readers as much time as possible to deal with it effectively. The progress report is part of the official documentation of the project. You don't want to be suspected of having covered up negative information.

Figure 8-4 is the conclusion for the computer-testing-device progress report. Note that because the design of the testing device will affect the future design of the new computer systems, the writer has wisely decided to ask for technical assistance—from the Research and Development Department.

Appendixes

In the appendixes, include any supporting materials that you believe your reader might wish to consult: computations, printouts, schematics, diagrams, charts, tables, or a revised task schedule. Be sure to provide cross-references to these appendixes in the body of the report, so that the reader can consult them at the appropriate stage of the discussion.

Figure 8-5 is an updated task schedule. The writer has taken the original task schedule from the proposal and added cross-hatching to show the tasks that have already been completed.

CONCLUSION

Phase I of the project has been completed successfully and on schedule.

We hope to work out the basic schematic of the testing device within two weeks. The one aspect of Phase II that is giving us trouble is the question of versatility. Although we can equip our future systems with a uniform integrated circuit that can be wired to produce unique identifiers, this procedure will create future headaches for R&D. We have arranged to meet with R&D next week to discuss this problem.

Figure 8-4. Conclusion Section

Tasks as represented on proposal
Tasks completed
Define desired characteristics
Design the device
Manufacture the device
Test the device
Submit the report
Nov. 1 Dec. 1 Jan. 1 Feb. 1 Mar. 1 Apr. 1
Time

Figure 8.5. Updated Task Schedule Section

EXERCISES

1. Write a progress report describing the work you are doing on a major project, such as the one you described in Exercise 1 in Chapter 7.

2. The following progress report was written by an engineer working for the waste-resources department of a medium-size city. In an essay, evaluate the report from the points of view of clarity, completeness, and writing style.

To: Lawrence Seigel, Head
Department of Water Resources
City of Corinth

From: Walter Prentice, P.E.

Subject: The Future of Municipal Sludge Composting

Introduction

Sludge composting is a 21-day process by which wastewater sludge is converted into organic fertilizer which is aesthetically acceptable, pathogen-free, and easy to handle. Composted sludge can be used to improve soil structure, increase the soil's water retention, and provide nutrients for plant growth.

Discussion

Sludge composting is essentially a two-step process:

1. Aerated-Pile Composting
Dump trucks deliver the dewatered raw sludge to the compost site. Approximately 10 tons of sludge is dumped on a 25 yd³ bed of bulking agent (usually woodchips). A front-end loader mixes the sludge into the bulking agent. The mixture is then placed on a compost pad and covered with a blanket of unscreened compost 1 ft. thick. This layer is applied to insulate the sludgebulking agent for ambient temperatures and for preventing the escape of odors from the pile. The air and odors are sucked out of the bulking agent base by an aeration system of pipes under the compost pad. After three weeks, the sludge in the aerated compost pile is essentially free of pathogens and stabilized.

2. Drying, Screening, and Curing the Composted Sludge
After the aerated pile composting is completed, the pile is spread out and harrowed periodically until it is dry enough to screen. Screening is desirable, because it recovers 80 percent of the costly bulking agent for reuse with new sludge. The screened compost is stored for at least 30 days before being distributed for use. During the curing period, the compost continues to decompose, ensuring an odor- and pathogen-free product.

Future Work

The next step is to determine the cost of sludge composting. Sludge composting utilizing the aerated-compost-pile method is estimated to cost between $35 and $50 per dry ton ($35 for a 50-dry-ton per day operation,

$50 for a 10-dry-ton per day operation). These estimates include all facilities, equipment, and labor necessary to compost at a site separate from the treatment plant. Not included are the costs of sludge dewatering, transportation to and from the site, and runoff treatment. These additional factors can raise the cost of composting to $160 per dry ton.

The breakdown of the capital costs for composting follows:

1. Site development—one acre of land is required for every three dry tons per day capacity. Half of the site should be surfaced. Asphalt paving costs about $60,000 per acre. In addition, the site requires electricity for the aeration blowers.

2. Equipment—front-end loader, trucks, tractors, screens, blowers, and pipes are required.

3. Labor—labor represents between a third and a half of the operating costs. Labor is estimated to cost $6 per hour, with five weeks of paid sick or vacation time.

However, a potential market exists for compost. Although the high levels of certain heavy metals in the compost restricts its use in some cases, compost can be used by businesses such as nurseries, golf courses, landscaping, and surface mining. Transportation costs can be high, however.

Conclusion

This information comes from published reports by federal agencies and journal articles. Since the government banned ocean sludge dumping in 1981, composting has become a viable method of waste treatment, and much has been written about it.

Before we can reach a final decision about whether composting would be economically justifiable for Corinth, we must add our numbers to the costs above. This process should take about two more weeks, provided that we can get all of the information.

9

Completion Reports

A completion report is generally the written record of a project that began with a proposal (see Chapter 7). In many cases, progress reports (see Chapter 8) followed the proposal.

The Role of the Completion Report

The completion report has two main functions: immediate communication and reference.

Readers want to know what the writer learned. Sometimes, the readers need only results: the technical data the writer created or discovered. Sometimes, the readers also need conclusions: interpretations of those data. And sometimes, the readers need one more item: recommendations for further action.

After the readers have finished reading the report, it is filed. As a reference document, a report is useful in helping new employees understand existing systems, processes, and equipment. A report is also useful if the organization is considering a new project and wants to determine how that new project would affect the existing situation. And finally, a filed report is useful if the system breaks down; the completion report will be the first place to look in planning how to fix it.

Types of Completion Reports

Completion reports can be classified into two broad categories: *physical-research reports* and *feasibility reports*. A physical-research report is written about a project that involves substantial empirical research, whether it is carried out in a lab or in the field. You begin with a hypothesis, conduct experiments to test it, record your results, and determine whether your hypothesis is correct as it stands or whether it needs to be modified.

A feasibility report documents a study that attempts to evaluate at least two alternative courses of action. For example, should our company hire a programmer to write a program we need, or should we have an outside company write it for us? Should we expand our product line to include a new item?

A feasibility study can address questions of possibility. We would like to build a new rail line to link our warehouse and our retail outlet, but if we cannot raise the cash, the project is not possible at this time. A feasibility study can also consider questions of economic wisdom. Even if we can raise

the money to build the rail line, is it a wise thing to do? If we use up all our credit on this project, what other projects will have to be postponed or canceled? Is there a less expensive or less risky way to achieve the same goals?

The Body of a Completion Report

Because most completion reports are formal, they usually contain many of the report elements discussed in Chapter 6. This chapter will discuss the body of a completion report. The body typically contains an introduction, methods, and some or all of the following elements: results, conclusions, and recommendations.

Introduction

Many writers prefer to put off writing the introduction until they have completed the other elements of the body. They reason that, in drafting the methods and the findings, they will have to deviate from their outline, and they will thus have to revise the introduction anyway.

The introduction enables the readers to understand the technical discussion that follows. Usually, the introduction contains most or all of the following elements:

1. *A statement of the problem or opportunity that led to the project.* What was not working, or not working well, in the organization? What improvements in the operation of the organization could be considered if more information were known? For instance, a manufacturing organization might want to determine the kind and nature of customer complaints before deciding whether the quality-control system needs to be revised. It might be useful or even necessary here to include a few paragraphs of background to orient the readers.
2. *A statement of the purpose of the project.* What exactly was the project intended to accomplish? What information was it intended to gather or create, or what action was it intended to facilitate?
3. *A statement of the scope of the project.* What aspects of the problem or opportunity were included in the project, and what aspects were excluded? For example, a report on new microcomputers might be limited to those that cost less than $4000, or those that have at least 512K of memory.

4. *An explanation of the organization of the report.* Readers understand better if they know where you are going and why. Explain your organizational pattern so the readers are not surprised or puzzled.
5. *A review of the relevant literature.* Sometimes, the literature will be internal, consisting of reports and memos produced within the organization. Sometimes, the literature will be external, consisting of published articles or even books that help your readers understand the context of your work.

Figure 9-1 shows an introduction to the body of a completion report. The subject of the report (MacBride 1986) is an investigation to determine whether a retail store can reduce its heating bills and increase its sales by instituting an energy-management system. The writer was a student working for the store in its Engineering and Planning Department. The report is entitled "Recommendation for Minimizing the Steam Consumption and Regulating the Ambient Temperature at Bridgeport Department Store."

Methods

The methods are the technical tasks or procedures you carried out. For a physical-research report, the methods will closely resemble a lab report. If several research methods were available to you, explain why you chose one method over the others. The equipment and materials you used should either be listed before the description of the research or discussed in the description itself.

For a feasibility study, the methods section will probably be more involved. You want to describe what you did and justify your actions. If you sent out a questionnaire, you should explain your rationale for the questions you asked and describe any pretesting you did on the questions. Of course, you should include a copy of the questionnaire in an appendix. If you visited a site, explain why you chose that site, as well as what you did when you got there. If you studied secondary sources, explain why you chose those particular sources.

Figure 9-2 is the methods section of the report on the energy-management system at the department store.

Results

The results are the data that you observed, discovered, or created. You should present the results objectively, so that the readers can "experience" the methods just as you did. Save the interpretation of the results—the

Introduction

Bridgeport Department Store, built in 1912, is an expensive building to heat because of its high ceilings, large windows, and lack of wall insulation. Every increase in heating costs hits us harder than it hits our competitors in more modern facilities. Therefore, it is necessary for us to make sure we are using the most modern and cost-effective methods for regulating our store temperature.

Currently, we have an energy-management system, but it regulates only our electricity for lights and air conditioning, not our steam used for heating.

In 1985, Bridgeport experienced a 6-percent increase in steam consumption over 1984 figures. This increase, combined with a 7-percent utility boost, resulted in a 15-percent ($67,000) rise in related operating expenses.

In addition, the number of customer complaints relating to the uneven temperatures throughout the building rose by 135 percent over the 1984 figures.

The purpose of the study is to determine whether it would be cost-effective to expand the capabilities of our current energy-management system to include the monitoring and regulating of steam consumption.

This study was restricted to changes in software. We do not have the funds to finance hardware changes to our main computer system, which was installed earlier this year.

After a brief discussion of the methods used in this study, the results are presented. The results section of the report describes how well the enhancement would meet the technical, management/maintenance, and financial criteria we established at the outset.

Recent articles by Cottrell and by Gorham on energy management indicate that an effective energy-management system can reduce steam usage by 20 to 25 percent. Donovan and Rossiter, writing on the relationship between store climate and consumer-buying behavior, suggest that the more comfortable the shoppers, the longer they stay and the more money they spend. This literature is discussed in the relevant sections of the report.

Figure 9-1. Introduction Section

Methods

Before studying the possible enhancement to our current energy-management system, we devised a set of technical, management/maintenance, and financial criteria by which to evaluate the enhancement. The results section of the report describes how well the enhancement would meet the appropriate criteria.

In performing the visual inspection of our system, we relied on internal records on energy usage and on blueprints, plans, and installation and operating manuals for our existing system. This literature is discussed in the relevant sections of the report. All the sources used in the report are listed in the bibliography, page 11.

Figure 9-2. Methods Section

conclusion—for later. If you intermix results and conclusions, your readers might be unable to follow your reasoning process. Consequently, they will not be able to tell whether the results justify the conclusions. Or your readers might think that you are slanting the results so that they lead to the conclusion you wanted all along.

In presenting the results, be specific. Providing full data, along with any necessary citations, enables your readers to do their own research if they wish. In addition, it gives them confidence in your professionalism.

A portion of the results section of the energy-management system report is shown in Figure 9-3.

Conclusions

The conclusions are the implications—the "meaning" of the results. It is important to state your conclusions clearly because many of your readers will not be able to interpret the data of your results. If, for instance, your results are that the exhaust emitted by the smokestacks in your plant contains three parts per million of sulfur dioxide, most of your readers will not know what that means. Is that a lot? Is it within EPA limitations? In writing your conclusions, you are trying to describe the situation as clearly as you can, so that the next section, the recommendations, will flow logically and inevitably.

The conclusion of the energy-management system report is shown in Figure 9-4.

Results

This section discusses the major technical components needed in an efficient energy-management system, and the tasks involved in converting our current system.

Major Technical Components of an Efficient Energy-Management System

There are two main technical components to an efficient energy-management system: a network of sensors and customized software.

SENSORS

Because climate is a function of humidity, entropy, and temperature levels, a network of sensors would have to be installed at strategic points throughout the building. These sensors would form the communication link to and from a particular area. Our energy-management system would continuously measure humidity, entropy, temperature, and steam-demand levels during a standard time interval. These actual climate measurements would be compared with predetermined target levels every few minutes. On the basis of these comparisons, the system would initiate the appropriate action (open or close the steam valves) to maintain the target temperature.

SOFTWARE

A computer program is needed to process the signals from the sensors, determine whether adjustments are required, and transmit instructions. This program would compare the parameter readings to the predetermined target levels, initiate the necessary adjustments (open or close the steam valves) for the target climate to be maintained, and record these measurements as data. These data could then be used to analyze current steam efficiency and potential energy-management opportunities.

Figure 9-3. Results Section

Recommendations

Recommendations are suggestions for future courses of action. A single recommendation is usually expressed in traditional sentences and paragraphs. Multiple recommendations are usually expressed in a numbered list.

State your recommendations clearly and tactfully. Remember that when you recommend a new course of action, you might be indirectly discrediting

Tasks Involved in Converting Our System

As discussed in the system description, a sophisticated energy-management system is based on a set of sensors and customized software. This section discusses the results of our analysis of the tasks required to convert our system to a sophisticated, computerized energy-management system.

SENSORS

One complication is that the current system employs steam valves that are operated by a pneumatic signal system. The system works on the principle of air pressure against a diaphragm; the expansion and contraction of the diaphragm opens and closes the valve. However, a sophisticated energy-management system can recognize only digital signals. Therefore, pneumatic-to-digital converters would have to be purchased and installed so that our system could regulate steam flow throughout the building. These converters could be installed by our electricians. The cost of the units, installed, would be less than $3500. See Appendix C, page 13, for descriptions and cost figures for the converters.

SOFTWARE

The program required to monitor and regulate the steam would be quite similar to the current program used to monitor and regulate the air-conditioning units throughout the building. For this reason, developing the steam-regulating program would require only about four person-weeks of work by our programming staff. This would cost less than $2000. In addition, several of the subprograms used by the air-conditioning system could also be used directly by the steam-heating program. This would reduce the amount of computer memory space needed. Therefore, the memory space currently available would be sufficient. See Appendix D, page 16, for the details of the programming needs.

Figure 9-3. Results Section *(Continued)*

a previous course of action. Therefore, you might run the risk of offending whoever formulated that earlier action. Don't write that your recommended action will "correct previous mistakes." Rather, write that your plan "offers great promise for success."

Figure 9-5 shows the recommendation section of the energy-management system report.

Conclusion

Incorporating our current steam-delivery system into an efficientenergy-management system would cost less than $6000 and could be accomplished in less than two weeks without our having to employ any additional personnel. This action would be the most cost-effective way to contain rising energy costs and improve the ambient temperature in the Bridgeport Department Store.

Figure 9-4. Conclusion Section

This report delivers good news: a relatively small expenditure will save the company a lot of money. Often, however, feasibility reports such as this one yield mixed news or bad news. That is, none of the available options would be an unqualified success, or none of the options would work at all. Don't feel that a negative recommendation reflects negatively on you. If the problem being studied were easy to solve, it probably would have been solved before you came along. Give the best advice you can, even if that advice is to do nothing. The last thing you want to do is to recommend a course of action that will not live up to the organization's expectations.

Recommendation

We recommend that Bridgeport Department Store incorporate the steam-delivery system into its existing energy-management system as soon as practical. The expenditure would pay for itself in less than one month. The Engineering and Planning Department stands ready to undertake this task.

Figure 9-5. Recommendation Section

EXERCISES

1. Write the completion report for the major assignment you proposed in Exercise 1 in Chapter 7.

2. The following table of contents and body are excerpted from a completion report written by a college student majoring in nutrition and addressed to a local school board (Desiata 1986). In an essay, evaluate the effectiveness of the excerpt from the points of view of clarity, completeness, and expression.

CONTENTS

II. INVESTIGATING NUTRITION EDUCATION IN SECONDARY EDUCATION AS AN AID IN THE PREVENTION OF EATING DISORDERS

A. Introduction

In the past eight years the incidence of eating disorders has increased by 7 percent among young adults. This increase means approximately 2.3 to 2.5 million more young adults are suffering from eating disorders than were in 1980. This report describes the investigation into the use of a nutrition-education course in the high school to decrease the incidence of eating disorders. To carry out this investigation, I conducted interviews with a nutrition specialist and a psychiatrist, administered a questionnaire to high-school students, and examined the school budget. I analyzed the information from the above sources against technical and financial criteria.

B. Course Criteria

In the effort to determine all the materials that should be covered in a nutrition education course, I spoke with Dr. Robert Abrams, R.D., a nutrition specialist, Anne McCoy, M.D.,a psychiatrist; and the high-school students of four area high schools. From these conversations, my focus was to determine: (1) the need for a nutrition education course, (2) the extent of the student's nutritional knowledge, (3) the acceptance of a nutrition education course by the students.

1. The Need for a Nutrition Education Course

The most important aspect of offering a course such as this is to determine if the course is, in fact, to the student's benefit. At a recent United Nations meeting with FAO (Food and Agricultural Organization) and WHO (World Health Organization) participation, it was said: "Nutrition education is a new and important concept which has a definite role to play in education. Although in many cases the new role can be evolved gradually without overloading the school curriculum, there is a desperate need for nutrition education, and foods and nutrition may have to be given priority over some other subjects already being taught." There are many advantages to teaching nutrition education to school-aged children. They tend to be more open-minded than adults and more used to accepting new knowledge and new habits as a part of growing up.

As part of my investigation I spoke with Dr. Robert Abrams, R.D., a nutrition specialist at Eastern University; our conversation revealed Dr. Abrams's belief that children aged from 10 to 16 need some introduction to nutrition education. Dr. Abrams stated, "The children of today are so used to convenience foods that they have no real concept of proper nutritional practices. They base most of what they learn about nutrition on what they see on television and in magazine advertisements. They need some formal training on the subject of nutrition." Dr. Anne McCoy, a psychiatrist who deals with patients suffering from eating disorders, stated that "eating disorders have reached epidemic levels and have become a real social concern."

2. The Extent of the Student's Nutritional Knowledge

To determine the extent of the average high-school student's nutritional knowledge, I administered a questionnaire with multiple-choice nutritional questions to the students of four area high schools. The high schools I chose for the distribution of this questionnaire all have comparable populations and have curricula that do not offer a nutrition education course. The schools I chose are Harrison High School, Bayshore High School, St. Patrick's High School, and Brentwood High School. The questionnaire test results for these four schools are:

School	*Average Score*
Harrison High School	64%
Bayshore High School	57%
St. Patrick's High School	61%
Brentwood High School	59%

The average grade for all the high schools is only a 60 percent, a grade that is below the accepted passing grade of 65 percent. This is an indicator that our students are not fully aware of even the basics of nutritional practices. For an example of some of the nutritional questions, refer to Appendix A.

3. The Acceptance of a Nutrition Education Course by the Students

The success of any course depends on the acceptance of the course by the students to whom it is being offered. The questionnaire that was administered to the students also proved as a medium to determine the acceptance of the course by the students.

The students were asked, in the questionnaire, whether they would be interested in participating in a nutrition education course if it were offered to them. (For a sample of these questions, refer to Appendix A under the course acceptance section.) The questionnaire results received (in this section) for the four high schools are:

School	*% Acceptance*
Harrison High School	85%
Bayshore High School	79%
St. Patrick's High School	75%
Brentwood High School	89%

The average acceptance percentage for the four schools is 82 percent. This average percentage indicates that the majority of the students would be interested in participating in a nutrition education course if it were offered in their curriculum.

C. Technical Criteria

Whenever a new course of study is offered into a curriculum, there are certain technical characteristics that must be analyzed. For the proposed nutrition education course, the technical criteria that must be analyzed are (1) teacher certification requirements, (2) course audience, and (3) course emphasis. All these technical criteria must be analyzed to enable an educator to make up the best possible course that will reach the targeted audience.

1. Teacher Certification Requirements

In New York State, the certification requirements for a teacher to teach a nutrition education course are not very stringent. The requirements are that the educator:

1. be certified to teach in New York State
2. have a bachelor of science degree in nutrition education, home economics, dietetics, or biology
3. attend a minimum of five approved nutrition conventions every three years.

These criteria would indicate that there are two teachers presently on the staff of the high school that would be eligible to teach a nutrition course. These two teachers are Audrey Rodgers, M.S., R.D., the home-economics teacher, and George Johnson, M.S., the biological science teacher.

2. Course Audience

Directing a course to a certain audience requires that a degree of restraint be placed on the materials covered and the way in which they are presented. To do this, the typical characteristics of the audience must be examined.

For the proposed nutrition education course, the audience will be high-school students that we must assume have a minimal amount of nutritional knowledge. This requires that the course material be presented in rather simple fashion. The terms used should be simple, easy to understand, and presented clearly. The extent of the student's biological science background should also be taken into account, and highly scientific terms and explanations should be avoided. However, the teacher should be able to help the student who shows an interest in a more in-depth understanding of a particular topic. It is important that the teacher try to reach all levels of intelligence in a classroom, without insulting any student's intelligence.

3. Course Content

Specialists agree that the course content for a high-school nutrition education course include all the basics of nutrition. On top of these basic concepts, equal time must also be given to improper dietary practices such as fad diets and eating disorders. Topics to discuss should be as follows:

a. the four basic food groups
b. vitamins and minerals; their importance and function
c. the dietary exchange system
d. diet management practices
e. fad diets
f. eating disorders; the causes and effects.

These topics should be presented, as stated previously, in a simple and clearly understandable manner.

D. Financial Criterion

The proposed nutrition course will be considered an elective course. Therefore, it is subject to approval by the Board of Educa

tion for admittance into the curriculum along with financial allocation in the budget.

After reviewing the budget for the year 1986–1987, I found that the allocation of money for an elective course initially is \$4875.00. This amount of money would be more than adequate for the institution of a nutrition education course into our curriculum. Accounting for textbooks, informational supplements such as pamphlets and films, and teacher's salary, the amount of money that will be needed for this nutrition education course, initially, is \$4500.00. This figure is \$375.00 below budget, and this savings may be used elsewhere as needed.

E. Evaluation Techniques

To evaluate the long-term effects of a nutrition education course on the incidence of eating disorders would require a long-term study of the lives of the students that took part in the course. This may be a hard procedure for our school to perform, because we lack the funding and expertise for such a study.

However, we may be able to predict what our success rate would be against predetermined data on this subject. Any institution that has an eating-disorder program will also offer nutrition education to these patients. It is from this practice that we can determine how well nutrition education helps the troubled eater. It is estimated by the Graduate Hospital located in Philadelphia, Pennsylvania, that the long-term success of cured eating-disorder patients depends approximately 46 percent on their acceptance of the nutritional education information that is presented to them. Specialists believe that this is because the patients now have a clear understanding of how important food is to their bodies and how important it is to feel good about oneself. The focus of this proposed nutrition education course is to try to get this point across to the students. What better way than to educate them on the importance of proper nutrition?

F. Conclusion

Early influences in an individual's life have profound and lasting effects not only on emotional and mental characteristics but also on anatomical and physiological characteristics as an adult.

Eating disorders are a problem that have reached epidemic levels in our society. There is such a push in our country to be perfect that many people have lost sight of proper nutrition. Too many people are victims of the "get thin quick" and even "gain weight fast" gimmicks on the market.

In order that we may lower this eating disorder incidence, we must educate the people. The best way to do this is to educate them when they are young and impressionable enough to accept this knowledge. This, in turn, will require our schools to institute programs that will do this.

There are many subjects of nutrition that should be covered by these programs. Correct nutritional practices, correct diet procedures, and the many aspects of eating disorders are among the subjects that should be presented to the students of today, but the main thrust should be on emphasizing a healthy body as part of a healthy mind.

G. Recommendation

I recommend that we institute a nutrition education program into the Islip High School as an aid in the prevention of eating disorders.

6

References

Desiata, T. 1986. Recommendation for the introduction of a nutrition education course into the secondary education curriculum as an aid in the prevention of eating disorders. Unpublished document.

MacBride, M. 1986. Recommendation for minimizing the steam consumption and regulating the ambient temperature at Bridgeport Department Store. Unpublished document.

10

Oral Presentations

An oral technical presentation is the spoken form of technical writing. Although there is, of course, a big difference between the crafts of writing and speaking, there are also many similarities. Most of the skills used in writing are used in preparing an oral presentation.

This chapter will concentrate on two kinds of oral presentations: extemporaneous and scripted. An extemporaneous presentation is planned in advance, but the speaker refers to notes and creates the sentences as the talk proceeds. A scripted presentation is written out in advance, and the speaker reads from a script. The extemporaneous presentation is preferred for most situations because, at its best, it is clear and spontaneous. A scripted presentation is often preferred on formal occasions because it increases the clarity and precision of the information; however, scripted presentations often sound stilted.

Two other kinds of oral presentations should be mentioned. An impromptu presentation is not planned in advance. The speaker is simply asked to talk. Rarely will an important topic be addressed in an impromptu presentation. Similarly, a memorized presentation is rare in technical contexts, because few people can remember the necessary details of the subject matter.

The Role of Oral Presentations

Oral presentations have an advantage over written presentations: the speaker and the audience can talk to each other. The listeners can ask questions and provide alternative viewpoints. You can expect to give oral presentations to three different kinds of audiences on the job: clients and customers, colleagues within your organization, and fellow professionals.

Selling your product or service often begins with an oral presentation to a potential client's purchasing agents. They want to know why what you are offering beats the competition. When you make the sale, you will find yourself making presentations to groups of the client's employees. Within your own organization, you will make many different kinds of presentations. If you are the house expert on a particular subject, you will present it to subordinates and supervisors. If you attend an important conference or return from working on an out-of-town project, you will present the important findings. If you have a good idea about how to improve procedures at your organization, you might be asked to make an oral presentation as part of the proposal process. Presentations to fellow professionals occur at conferences and meetings.

The more advanced you are in your organization and in your field, the more you can expect to make oral presentations. But even new employees speak often. You probably have not had much experience at public speaking. However, making an oral presentation is a skill that, like technical writing, can be practiced and learned.

Preparing the Presentation

The first step in planning any written or oral presentation is to analyze your audience and purpose. Who are your listeners? What do they already know about the subject? What do they need to know? What is their attitude toward the subject? Are you trying merely to inform your audience or to change their attitudes? How will your audience and purpose affect the structure and content of the presentation?

As you think about these issues, don't forget the time limitation for your presentation. As a rough guide, most speakers require almost a minute to deliver effectively a double-spaced page of text. It is best to be brief so that you have time to answer questions. If you take more than your allotted time, you are being rude, not only to your audience but also to any other speakers.

Once you have analyzed your audience and purpose, begin preparing the notes.

Preparing the Notes

Almost all extemporaneous speakers need to refer to notes during their presentations. Start by writing an outline just as you would for a written presentation. As you rehearse the presentation, you will refine the outline. Finally, you will transfer it to note cards or to whatever medium you are most comfortable with.

Keep in mind that because your listeners cannot "reread" what you have said, an oral presentation should be somewhat less complex than a written one. First brainstorm, and then work on outlining the body of the presentation. Use any of the patterns of development that you would use in writing. Then, outline the conclusion. Make sure it emphasizes the main points and summarizes the presentation. Finally, go back and outline the introduction.

The introduction is crucial, because you must gain and keep the audience's attention. Begin with the problem or opportunity that led to the project you are discussing, or with an interesting fact the audience is unlikely to know. Sometimes an interesting quotation is useful. Forecast the major points of the presentation explicitly. Don't be fancy. Use the words *scope* and *purpose.* Don't try to spice up the presentation with a joke. Humor is usually inappropriate in professional presentations.

Preparing the Graphic Aids

Graphic aids fulfill the same purpose in an oral presentation that they do in a written one: they clarify or highlight important ideas or facts. Graphic aids should be clear and self-explanatory. The audience should know immediately what each graphic aid is showing. In addition, the graphic should illustrate a single idea. Remember that your listeners have not seen the graphic aid before and that they, unlike readers, do not have the opportunity to study it.

If possible, carefully consider the room in which you will be making the presentation. The people in the last row and near the sides of the room must be able to see each graphic aid easily. If you make a transparency from a page of text, for example, be sure to make the picture or words larger. What is legible on a printed page is usually too small to see on a screen.

Check and recheck your graphic aids for accuracy and correctness. A misspelled word or a number that is off by a factor of 10 is particularly embarrassing when it is visible to a whole roomful of people.

Although there are no firm guidelines on how many graphic aids to create, a good rule of thumb is to have a different one for every 30 seconds of the presentation. Changing from one to another helps keep the presentation visually interesting, and it helps you signal transitions to your audience. It is far better to have a series of simple graphics than to have one complicated one that stays on the screen for 10 minutes.

Following is a list of some of the basic media for graphic aids, with the major advantages and disadvantages cited:

1. *Overhead projector*: projects a transparency onto a screen.
   ADVANTAGES:
   Transparencies are inexpensive and easy to draw.
   You can create transparencies "live."
   You can create overlays by placing one transparency over another.
   You can leave the lights on during the presentation.
   You can face the audience.
   DISADVANTAGES:
   The appearance can be informal.
   You must load each transparency by hand.

2. *Opaque projector*: projects a piece of paper onto a screen.
   ADVANTAGES:
   It can project single sheets or pages in a bound volume.
   It requires no expense or advanced preparation.
   DISADVANTAGES:
   The room has to be kept dark.
   The projector cannot magnify sufficiently for a large audience.
   You must load each page separately by hand.
   The projector is noisy.
3. *Chalkboard.*
   ADVANTAGES:
   It is almost universally available.
   You have complete control; you can add, delete, and modify information.
   DISADVANTAGES:
   Complicated or extensive graphics are difficult to create.
   Chalkboards are ineffective in large rooms.
   Chalkboards have a very informal appearance.
4. *Objects*: models or samples of materials that can be held up or passed around through the audience.
   ADVANTAGES:
   They are very interesting for the audience.
   They provide a very good look at the object.
   DISADVANTAGES:
   Audience members might not be listening to you while they look at the object.
   Objects might not survive intact.
5. *Handouts*: photocopies of written material given to each audience member.
   ADVANTAGES:
   Much material can be fit on paper.
   Audience members can write on the paper and keep it.
   DISADVANTAGE:
   Audience members might read the paper rather than listen to you.

Before you design and create any graphic aids, make sure the room in which you will be giving the presentation has the equipment you need. On the day of the presentation, check to make sure the equipment is there and operating, even if you have arranged beforehand to have the equipment delivered. If possible, bring your own equipment.

Rehearsing the Presentation

Set aside enough time to practice your presentation three or four times. For the first rehearsal, simply sit at a desk, with your outline or notes before you. Try to talk through the presentation without worrying about voice projection or posture. Your goal here is to see whether the presentation makes sense: do you understand all the material, and can you forge transitions between one point and the next? Note the items that you need to research more. After you have gotten the additional information, revise your outline or notes.

After you have rested, try it again. This time, the information should flow more smoothly. Revise the information again, and keep practicing until you are confident about the material. Then, time it to make sure that you are within the time guidelines.

Finally, try the presentation under more realistic conditions. If you can get people to listen and offer suggestions, that's the best method. Otherwise, use a tape recorder—or, better yet, a video recorder—and evaluate yourself. If possible, visit the site of the presentation to become comfortable with the environment.

For a scripted presentation, use the same rehearsal technique: try to get people to listen to you and to evaluate [illegible]

Delivering the Presentation

Most professional performers are nervous before a performance, so there is no reason that you shouldn't be. And remember that any signs of nervousness are much more apparent to yourself than they are to the audience. After you have completed the first minute or two of the presentation, you will probably begin to relax and to concentrate on your subject.

When it's time to give the presentation, walk up to the lectern and arrange your notes. Don't immediately start talking. Give yourself a few seconds to get ready. It is polite to begin formal presentations with "Good morning" (or "Good afternoon," etc.) and to include a reference to the officers and dignitaries present. If your name has not been mentioned by the introductory speaker, identify yourself. In less formal contexts, just begin your presentation.

Try to project the same image that you would in a job interview—that is, one of restrained self-confidence. Show your audience that you are interested in your topic and know what you're talking about. In giving the

presentation, this sense of control is achieved chiefly through your voice and your body.

Your Voice

Inexperienced speakers often encounter problems with five aspects of vocalizing.

1. *Volume.* Most people speak too softly in presentations. After a few sentences, ask if the people in the back of the room can hear you. If you have a microphone, don't bend down and talk right into it. Instead, hold it about a foot from your mouth and speak at a normal volume.
2. *Speed.* Speak more slowly than you normally do. Remember, your audience has not heard what you have to say. They need time to absorb it.
3. *Pitch.* Try to let the natural pitch variations of conversation come through. In fact, most experienced speakers exaggerate these variations, especially in bigger rooms.
4. *Clarity of pronunciation.* The nervousness that goes along with oral presentations often accentuates sloppy [illegible]. If [illegible] any [illegible]. Be careful when you introduce important terms. Say them slowly, even though *you* are totally familiar with them.
5. *Verbal fillers.* Avoid such meaningless fillers as "you know," "okay," and so forth. A slight pause is better than an annoying verbal tic.

Your Body

Your audience will be listening to you, but they will also be looking at you. Try to take advantage of this.

Eye contact is crucial. Look at your listeners. Scan the audience randomly. If you look at the floor or out the window, your audience will infer that you don't know your subject or that you are hiding something. Eye contact suggests sincerity and conviction. In addition, eye contact gives you important information about how well the audience can hear you.

Your arms and hands are also important. Watch experienced speakers use their arms and hands to signal pauses and emphasize important information. Gestures are particularly useful when you are referring to graphic aids. Use your arm to direct the audience's attention to an area of the graphic.

Try to avoid distracting mannerisms, such as pulling up your pants or compulsively scratching your head. Like verbal mannerisms, physical mannerisms are often unconscious; constructive criticism from friends can help you to pinpoint and eliminate them.

After the Presentation

Most oral presentations are followed by a question-and-answer period. To invite questions, say something such as, "If you have any questions, I'd be happy to try to answer them now." Then give them three to five seconds. If asked politely, people will be much more likely to ask questions; therefore, you will be much more likely to communicate your information effectively.

To make sure everyone has heard the question, repeat it for the audience. And make sure you understand it. Ask the questioner for clarification if necessary. Everyone's time is wasted if you answer the wrong question.

If you don't know the answer, say so. If possible, tell the questioner where the answer might be found, or offer to find it yourself and get in touch with him or her later.

After the question-and-answer part of the presentation has concluded, thank the audience for their courtesy in listening to you. If it is appropriate to stay after the session to talk individually with audience members, offer to do so.

11

Job-Application Materials

Job-application materials are crucial in an obvious way: potential employers want to know as much as they can about your background and experience in order to decide whether to interview you. However, job-application materials are equally important in another, more subtle way: they give readers a very good idea of your communication skills, as well as of your work habits, diligence, maturity, and so forth. The care with which you prepare these materials is every bit as important as the information contained in them. If there is any occasion to do a perfect job communicating, applying for a job is it.

Whether you are writing in response to a published job ad or to an organization that has not advertised at all, the process is basically the same. First, you find out everything you can about the organization. Then, you compose a résumé and a job-application letter. If you then get to interview with the organization, you write a follow-up letter.

Start by researching the company. What does the job advertisement say they are looking for? Consult the organization's annual report, if it is a corporation. The business librarian at your college can help you find other sources of information, such as the Dun and Bradstreet guides, the *F & S Index of Corporations*, and the indexed newspapers such as the *New York Times* and the *Wall Street Journal*.

If you are writing because someone you know has indicated that a job might exist, talk in detail with this person to find out about the company's current and future projects, priorities, and policies.

The Résumé

Your job-application materials communicate by their appearance and their content. Readers notice the appearance first.

You can have your résumé professionally typeset and printed, or you can merely type it neatly and photocopy it on good paper. Studies suggest that it doesn't make much difference. But it is essential that it look professional. Use one-inch margins on all four sides. Use a typewriter or word processor with clear, unbroken type; do not use a dot-matrix printer. Create a symmetrical, balanced appearance. Make the organization of the résumé clear. The reader should be able to tell at a glance where one section of the résumé begins and ends; white space *between* items should be greater than that *within* items. And, of course, the résumé must be error free; spelling errors and typos suggest a casual attitude toward your work.

The résumé must provide specific information. Avoid generalizations.

You cannot say that you are a terrific job candidate, as if you were selling a car. But you can provide the facts that lead the reader to the impression that you are a terrific candidate. Show your readers; don't tell them.

Most students' résumés are one page. If, however, you have a lot of good information, such as previous work experience, publications, or patents, you can go to a second page. But don't pad the résumé with unimportant information to make it longer. That suggests poor judgment.

Most résumés contain at least five basic sections:

1. *Identifying information.* Your name, address, and phone number (including your area code) should appear prominently at the top of the page.
2. *Education.* List at least the name and city of the institution, the degree you expect, and the date of graduation. Also provide any other information that will fill out the section: (1) your grade-point average (overall or in your major, or both) if it is substantially above the average; (2) lists of important, unusual, or advanced courses you have taken; (3) lists of academic honors you have earned; and (4) descriptions of special projects you have carried out. Use the education section to suggest that you used your time at college constructively. Distinguish yourself from the other students in your major who might be competing for the same job.

   The education section comes after the identifying information for most students. If you have substantial professional work experience, the employment section might precede the education section.
3. *Employment experience.* For each professional-level job you have had, list the dates of employment, the name and location of the organization, and your position or title. Then describe what you did. Focus on the kinds of skills and experience your potential employer will find valuable: kinds of reports you wrote, numbers and kinds of clients you served, skills you applied or acquired, kinds of equipment or machinery you operated, amounts of money you were responsible for, numbers of people you supervised. In your description, use verbs that show responsibility, such as *administered, analyzed, constructed, delivered, developed, directed, edited, evaluated, hired, improved, increased, managed, operated, organized, prepared, purchased, reported, supervised,* and *wrote*.

   Generally, jobs are sequenced in reverse chronological order. If you have held a number of nonprofessional jobs (such as waitress or clerk) in addition to several professional jobs, group the nonprofessional ones together at the end of the section in a format such as the following:

   Other employment: Cashier (summer 1986), clerk (summer 1987).

In this way, these positions will not distract the reader's attention from the more professional jobs.

4. *Personal information.* Most personnel officers do not want to see traditional personal information such as height and weight. But this section is the place for miscellaneous information, such as participation in extracurricular activities at school, volunteer work, and hobbies that are related to your professional goals. This miscellaneous information shows that you are an active, well-rounded applicant. Omit items that might create a negative impression in some employers, such as hunting, gambling, performing in a rock band, and so forth. And omit such activities as reading and meeting people—everyone does these things.
5. *References.* It is best to list three or four referees—people who are willing to vouch for you. Listing the referees gives the impression that you are completely prepared for the application process and that you have nothing to hide. For each referee, list the name, position, mailing address, and phone number.

   Choose people who know your work well at school and on the job. Omit character referees, such as clergy who cannot speak about your professional credentials. When you ask if a person is willing to be a referee, give him or her the opportunity to decline gracefully. Many potential referees are too embarrassed to say, "No, I don't think that highly of you." Therefore, you have to phrase the question carefully. For example, you might ask, "Do you think you know me well enough to write a very positive letter?"

Other sections that sometimes appear in résumés cover such credentials as service in the armed forces, honors, and knowledge of a foreign language.

One other common element of résumés should be mentioned here: the objective section ("Objective: an entry-level position in civil engineering, leading to management opportunities"). Most employment officers say it is of little value because most students don't know enough about the field they wish to enter to say with any specificity what their objectives are. In fact, the objective statement often works against the candidate because it can indicate to the potential employer that the candidate would not be happy at the organization.

Figure 11-1 is an example of an effective résumé.

The Job-Application Letter

The job-application letter is the first thing a potential employer sees. Its function is to highlight and elaborate on several items from the résumé. Most people do not have the patience to make up different résumés for each

KENNETH CHAING
(215) 525-6881

753 Westborn Drive
Ardmore, PA 19316

EDUCATION

B.S. in Civil Engineering
Eastern University, Lynwood, PA
Anticipated June 1988

Grade-Point Average: 3.35 (of 4.0)

Advanced Business Courses

Financial Accounting
Legal Options in Decision Making
Manpower Management
Budgeting
Advanced Accounting
Labor Law

EMPLOYMENT

May 1987-September 1987
Gilmore Construction, Redford, PA
Prepared bids and estimates for storm and sanitary piping.
Revised drafting for 700-unit housing complex.
Ordered piping materials amounting to $325,000.

June 1986-September 1986
Gilmore Construction, Redford, PA
Worked in drafting and reproductions department.
Supervised piping sales.

June 1985-September 1985
Pertwell Construction, Salford, PA
Laborer.

PERSONAL INFORMATION

Member, American Society of Civil Engineers
Eagle Scout

REFERENCES

Mr. Allen Chrome
President
Gilmore Construction
Frazer Park
Redford, PA 18611
(306) 912-1773

Mr. Len Lefkowitz, P.E.
Chief Engineer, Pertwell Construction
1911 Market Street
Redford, PA 18611
(306) 432-1814

Dr. Harold Murphy
Professor of Civil Engineering
Eastern University
Lynwood, PA 19314
(215) 669-4300

Figure 11-1. Résumé

potential employer; therefore, the letter has to address the needs of the particular reader.

If the letter does not make a good impression, the reader probably will not bother to read the résumé. This is a challenge because each letter must be typed (or word-processed) individually. By contrast, the résumé is easier because it has to be perfect only once. Many employment officers admit that one of the things they look for first is the neatness of the letter. They check it for formatting, typos, strikeovers, corrections, and so on. Their thinking is that if you really want the job, you'll make sure the letter looks truly professional.

Although each job-application letter is unique, a common approach is to use a four-paragraph structure:

1. *Introduction.* The introductory paragraph should identify how you found out about the vacancy and should specify for which job you wish to be considered. Often, an organization will have a number of job ads circulating at one time; unless you specify, the reader might not know which position you seek. In addition, your introductory paragraph should explicitly state that you wish to be considered for the position. And finally, you should include a few phrases perhaps one about your education and one about your experience—that forecast the body of the letter. You want to show your reader that you are not asking for a favor, but making the case that you can meet the organization's needs.
2. *Education.* Study the job ad carefully to see what the potential employer is looking for. Then determine what aspect of your education best responds to the requirements. Try to create a unified paragraph with a topic sentence. In some cases, you might describe a set of courses that you took that covers the areas mentioned. In other cases, an ambitious senior project you are working on might show the necessary kinds of skills and experience.
3. *Experience.* Again, try to create a unified paragraph. In general, one of two approaches works best for the employment paragraph; your strong suit will be either (1) a single position that matches the potential employer's specified needs or (2) the range and diversity of your experience. Regardless of the strategy you use in developing the paragraph, be sure to use plenty of specific details. Your reader will not be impressed with vague statements such as, "At Hobson Electronics I learned a great deal about the electronics industry." What exactly did you learn, and how will it help the potential employer?

753 Westborn Drive
Ardmore, PA 19316
November 23, 1988

Mr. Arnold Peck
Director of Personnel
Lientz Construction, Inc.
119 Westview Drive
Willoughby, OH 44094

Dear Mr. Peck:

With a civil engineering degree from Eastern University and practical experience in the field, I believe I could be of value to Lientz Construction. Would you please consider me for the junior civil engineering position described in the November 22 *New York Times?*

Your notice calls for a candidate with "business sense." While at Eastern, I took many business courses, including three advanced seminars. I was intrigued by the sometimes conflicting goals of high profits and good labor-management relations and did my research on solutions to such problems in the construction industry.

In three summers' work with two construction firms, I saw the practical side of this issue. I began as a laborer, experience which will be of great value to me in my career in construction. I have prepared over three dozen bids and estimates for both residential and commercial customers. In addition, I revised the entire drafting of a 700-unit housing complex in southern New Jersey.

The enclosed résumé provides an overview of my skills and experience. Could I meet with you at your convenience to discuss my qualifications for this position? You can leave a message for me any weekday at (215) 525-6681.

Very truly yours,

Kenneth Chaing

Kenneth Chaing

Enclosure (1)

Figure 11-2. Job-Application Letter

1901 Chestnut Street
Phoenix, AZ 63014

July 13, 19--

Mr. Daryl Weaver
Director of Operations
Cynergo, Inc.
Spokane, WA 92003

Dear Mr. Weaver:

I would like to thank you for taking the time yesterday to show me your facilities and to introduce me to your colleagues.

Your advances in piping design were particularly impressive. As a person with hands-on experience in piping design, I can fully appreciate the advantages your design will have.

The vitality of your projects and the obvious good fellowship among your employees further confirm my initial belief that Cynergo would be a fine place to work. I would look forward to joining your staff.

Sincerely yours,

Albert Rossman

Figure 11-3. Follow-up Letter

4. *Conclusion.* If you have not already mentioned that your résumé is enclosed, do so in the concluding paragraph. Also, mention your phone number and explicitly request an interview, "at your convenience."

Figure 11-2 is an example of an effective job-application letter. The letter accompanies the résumé that appears in Figure 11-1.

The Follow-up Letter

The final document in the job search is the follow-up letter, which is sent to a potential employer after an interview or plant tour. Your purpose is to thank the representative for taking the time to see you and to remind him or her of your particular qualifications for the job.

The follow-up letter can do more good with less effort than any other step in the job search. Why? Only about one in ten candidates takes the trouble to write it.

Figure 11-3 is an example of an effective follow-up letter.

EXERCISES

1. In an essay, evaluate the effectiveness of the following job-application letter. What changes would you make to improve it?

April 13, 1988

Wayne Grissert
Best Department Store
113 Hawthorn
Atlanta, Georgia

Dear Mr. Grissert:

As I was reading the Sunday *Examiner*, I came upon your ad for a buyer. I have always been interested in learning about the South, so would you consider my application?

I will receive my degree in fashion design in one month. I have taken many courses in fashion design, so I feel I have a strong background in the field.

Also, I have had extensive experience in retail work. For two summers I sold women's accessories at a local clothing store. In addition, I was a temporary department head for two weeks.

I have enclosed a résumé and would like to interview you at your convenience. I hope to see you in the near future. My phone number is 436-6103.

Sincerely,

Brenda Adamson

Brenda Adamson

THOMAS MARKS
2004 South Street
Philadelphia, PA 19138
(215) 895-2444

EDUCATION

Eastern University, Philadelphia, PA (9/84–12/87)
B.S. Mechanical Engineering, May 1988

EMPLOYMENT HISTORY

McManus Incorporated, Hatboro, PA (3/87–12/87)

Design Engineer – Responsible for design and development of automatic deflashing machinery for the plastics industry. Duties included design calculations, layout and detail drafting of machine parts and assemblies, component selection, and bill of material compilation. Served as shop/office liason. Machine shop experience includes milling, lathe work, welding, and sheet metal fabrication. Responsible for finding and solving design problems and correcting manufacturing errors.

COOPERATIVE EMPLOYMENT

Allied Machinery Company, Langhorne, PA (1/84–6/84
Consumer Packaging Division 1/85–6/85
Machinery Development Group 1/86–6/86)

Development Technician – Involved in many phases of packaging machinery design and development, from conception through layout and detail drafting, fabrication, assembly, testing, operation, and [illegible] for testing and modifying a variety of packaging equipment and systems in a proof of principle environment. Duties included drafting, machining, and assisting with customer applications and installations with approximately twenty percent travel.

PROFESSIONAL MEMBERSHIP

Student member of the American Society of Mechanical Engineers, American Institute of Aeronautics and astronautics.

REFERENCES

Furnished upon request

2. In an essay, evaluate the effectiveness of the accompanying résumé. What changes would you make to improve it?

3. In an essay, evaluate the effectiveness of this follow-up letter. What changes would you make to improve it?

123 Center St.
Camden, New Jersey

Robert Montoya
Personnel
SMC Corporation
Industrial Highway
Paoli, PA

Dear Mr. Montoya,

Thank you very much for letting me interview you last week. I hope that I made the case that I would fit right in in your company.

I will be ready to start work by mid May. I look forward to hearing from you.

Sincerely,

Howard Stoops

Howard Stoops

Appendix A

Handbook of Style, Punctuation, and Mechanics

This handbook concentrates on grammar, style, punctuation, and mechanics. Where appropriate, common errors are defined directly after the correct usage is discussed.

Many of the usage recommendations made here are only suggestions. If your organization or professional field has a style guide that makes different recommendations, you should of course follow it.

Also, note that this is a selective handbook. It cannot replace full-length treatments, such as the handbooks often used in composition courses.

Style

Use Modifiers Effectively

Technical writing is full of modifiers—phrases and clauses that describe other elements in the sentence. To make your meaning clear, you must use modifiers effectively. You must clearly communicate to your readers whether a modifier provides necessary information about its referent (the word or phrase it refers to) or whether it simply provides additional information. Furthermore, you must make sure that the referent itself is always clearly identified.

RESTRICTIVE AND NONRESTRICTIVE MODIFIERS

A **restrictive modifier,** as the term implies, restricts the meaning of its referent: that is, it provides information necessary to identify the referent. In the following example, the restrictive modifiers are italicized.

> **The aircraft *used in the exhibitions* are slightly modified.**
> **Please disregard the notice *you just received from us.***

In most cases, the restrictive modifier doesn't require a pronoun, such as *that* or *which.* If you choose to use a pronoun, however, use *that:* "The aircraft that are used in the exhibits are slightly modified." (If the pronoun refers to a person or persons, use *who.*) Note that restrictive modifiers are not set off by commas.

A **nonrestrictive modifier** does not restrict the meaning of its referent: in other words, it provides information that is not necessary to identify the referent. In the following examples, the nonrestrictive modifiers are italicized.

The personal computer market, *which didn't exist a decade ago,* is now estimated to be worth $10 billion annually.
When you arrive, go to the Registration Area, *which is located on the second floor.*

Like the restrictive modifier, the nonrestrictive modifier usually does not require a pronoun. If you use one, however, choose *which* (*who* or *whom* when referring to a person or persons). Note that nonrestrictive modifiers are separated from the rest of the sentence by commas.

MISPLACED MODIFIERS

The placement of the modifier often determines the meaning of the sentence. Note, for instance, how the placement of *only* changes the meaning in the following sentences.

Only Turner received a cost-of-living increase last year.
(*Meaning:* Nobody else received one.)

Turner received only a cost-of-living increase last year.
(*Meaning:* He didn't receive a merit increase.)

Turner received a cost-of-living increase only last year.
(*Meaning:* He received a cost-of-living increase as recently as last year.)

Turner received a cost-of-living increase last year only.
(*Meaning:* He received a cost-of-living increase in no other year.)

Misplaced modifiers—those that appear to modify the wrong referent—pose a common problem in technical writing. The solution is, in general, to make sure the modifier is placed as close as possible to its intended referent. Frequently, the misplaced modifier is a phrase or a clause:

MISPLACED

The subject of the meeting is the future of geothermal energy in the downtown Webster Hotel.

CORRECT

The subject of the meeting in the downtown Webster Hotel is the future of geothermal energy.

MISPLACED

Jumping around nervously in their cages, the researchers speculated on the health of the mice.

CORRECT

The researchers speculated on the health of the mice jumping around nervously in their cages.

A special kind of misplaced modifier is called a **squinting modifier**—one that is placed ambiguously between two potential referents, so that the reader cannot tell which one is being modified:

UNCLEAR

We decided immediately to purchase the new system.

CLEAR

We immediately decided to purchase the new system.

CLEAR

We decided to purchase the new system immediately.

UNCLEAR

The men who worked on the assembly line reluctantly picked up their last paychecks.

CLEAR

The men who worked reluctantly on the assembly line picked up their last paychecks.

CLEAR

The men who worked on the assembly line picked up their last paychecks reluctantly.

DANGLING MODIFIERS

A **dangling modifier** is one that has no referent in the sentence:

Searching for the correct answer to the problem, the instructions seemed unclear.

In this sentence, the person doing the searching has not been identified. To correct the problem, rewrite the sentence to put the clarifying information either *within* the modifier or *next to* the modifier:

As I was searching for the correct answer to the problem, the instructions seemed unclear.
Searching for the correct answer to the problem, I thought the instructions seemed unclear.

A writer sometimes can correct a dangling modifier by switching from the indicative mood (a statement of fact) to the imperative mood (a request or command):

DANGLING

To initiate the procedure, the BEGIN button should be pushed.

CORRECT

To initiate the procedure, push the BEGIN button.

In the imperative, the referent—in this case, *you*—is understood.

Keep Parallel Elements Parallel

A sentence is **parallel** if its coordinate elements are expressed in the same grammatical form: that is, all its clauses are either passive or active, all its verbs are either infinitives or participles, and so forth. Creating and sustaining a recognizable pattern for the reader makes the sentence easier to follow.

Note how faulty parallelism weakens the following sentences.

NONPARALLEL

Our present system is costing us profits and reduces our productivity. (*nonparallel verbs*)

PARALLEL

Our present system is costing us profits and reducing our productivity.

NONPARALLEL

The dignitaries watched the launch, and the crew was applauded. (*nonparallel voice*)

PARALLEL

The dignitaries watched the launch and applauded the crew.

NONPARALLEL

The typist should follow the printed directions; do not change the originator's work. (*nonparallel mood*)

PARALLEL

The typist should follow the printed directions and not change the originator's work.

A subtle form of faulty parallelism often occurs with the correlative constructions, such as *either . . . or, neither . . . nor*, and *not only . . . but also*:

NONPARALLEL

The new refrigerant not only decreases energy costs but also spoilage losses.

PARALLEL

The new refrigerant decreases not only energy costs but also spoilage losses.

In this example, *decreases* applies to both *energy costs* and *spoilage losses*. Therefore, *decreases* should precede the first half of the correlative construction. Note that if the sentence contains two different verbs, the first half of the correlative construction should precede the verb:

The new refrigerant not only decreases energy costs but also prolongs product freshness.

When creating parallel constructions, make sure that parallel items in a series do not overlap, thus changing or confusing the meaning of the sentence:

CONFUSING

The speakers will include partners of law firms, business executives, and civic leaders.

CLEAR

The speakers will include business executives, civic leaders, and partners of law firms.

The problem with the original sentence is that *partners* appears to apply to *business executives* and *civic leaders*. The revision solves the problem by rearranging the items so that *partners* cannot apply to the other two groups in the series.

Parallelism should be maintained not only within individual sentences but also among sentences in paragraphs. When you establish a pattern in a paragraph, follow it through. The pattern may be as simple as numbering the steps in a process:

PARALLEL

Correlating the two results is a three-part procedure. First, ... Second, ... And third, . . .

Keep the voice consistent—either active or passive:

PARALLEL

The sample is placed in the petri dish. Two drops of culture are added. . . . After two days, the growth rate is examined by. . . . Finally, the sample is weighed.

Avoid Ambiguous Pronoun Reference

Pronouns must refer clearly to the words or phrases they replace. **Ambiguous pronoun references** can lurk in even the most innocent-looking sentences:

UNCLEAR

Remove the cell cluster from the medium and analyze it. (*Analyze what, the cell cluster or the medium?*)

CLEAR

Analyze the cell cluster after removing it from the medium.

CLEAR

Analyze the medium after removing the cell cluster from it.

CLEAR

Remove the cell cluster from the medium. Then analyze the cell cluster.

CLEAR

Remove the cell cluster from the medium. Then analyze the medium.

Ambiguous references can also occur when a relative pronoun, such as *which*, or a subordinating conjunction, such as *where*, is used to introduce a dependent clause:

UNCLEAR

She decided to evaluate the program, which would take five months. (*What would take five months, the program or the evaluation?*)

CLEAR

She decided to evaluate the program, a process that would take five months. (*By replacing "which" with "a process that," the writer clearly indicates that the evaluation will take five months.*)

CLEAR

She decided to evaluate the five-month program. (*By using the adjective "five-month," the writer clearly indicates that the program will take five months.*)

UNCLEAR

This procedure will increase the handling of toxic materials outside the plant, where adequate safety measures can be taken. (*Where can adequate safety measures be taken, inside the plant or outside?*)

CLEAR

This procedure will increase the handling of toxic materials outside the plant. Because adequate safety measures can be taken only in the plant, the procedure poses risks.

CLEAR

This procedure will increase the handling of toxic materials outside the plant. Because adequate safety measures can be taken only outside the plant, the procedure will decrease safety risks.

As the last example shows, sometimes the best way to clarify an unclear pronoun is to split the sentence in two, eliminate the problem, and add clarifying information. Clarity is always the primary characteristic of good technical writing. If more words will make your writing clearer, use them.

Ambiguity can also occur at the beginning of a sentence:

UNCLEAR

Allophanate linkages are among the most important structural components of polyurethane elastomers. They act as cross-linking sites. (*What act as cross-linking sites, allophanate linkages or polyurethane elastomers?*)

CLEAR

Allophanate linkages, which are among the most important structural components of polyurethane elastomers, act as cross-linking sites. (*The writer has changed the second sentence into a clear nonrestrictive modifier.*)

Your job is to use whichever means—restructuring the sentence or dividing it in two—will best assure that the reader will know exactly which word or phrase the pronoun is replacing.

Compare Items Clearly

When comparing or contrasting items, make sure your sentence clearly communicates the relationship. A simple comparison between two items often causes no problems: "The X3000 has more storage than the X2500." However, don't let your reader confuse a comparison with a simple statement of fact. For example, in the sentence "Trout eat more than minnows," does the writer mean that trout don't restrict their diet to minnows or that trout eat more than minnows eat? If a comparison is intended, use a second verb: "Trout eat more than minnows do." And if three items are introduced, make sure that the reader can tell which two are being compared:

AMBIGUOUS

Trout eat more algae than minnows.

CLEAR

Trout eat more algae than they do minnows.

CLEAR

Trout eat more algae than minnows do.

Beware of comparisons in which different aspects of the two items are compared:

ILLOGICAL

The resistance of the copper wiring is lower than the tin wiring.

LOGICAL

The resistance of the copper wiring is lower than that of the tin wiring.

In the illogical construction, the writer contrasts *resistance* with *tin wiring* rather than the resistance of copper with the resistance of tin. In the revision, the pronoun *that* is used to substitute for the repetition of *resistance*.

Use Adjectives Clearly

In general, adjectives are placed before the nouns they modify: *the plastic washer*. Technical writing, however, often requires clusters of adjectives. To prevent confusion, use commas to separate coordinate adjectives, and use hyphens to link compound adjectives.

Adjectives that describe different aspects of the same noun are known as coordinate adjectives:

a plastic, locking washer

In this case, the comma replaces the word *and*.

Note that sometimes an adjective is considered part of the noun it describes: *electric drill*. When one adjective is added to *electric drill*, no comma is required: *a reversible electric drill*. The addition of two or more adjectives, however, creates the traditional coordinate construction: *a two-speed, reversible electric drill*.

The phrase *two-speed* is an example of a compound adjective—one made up of two or more words. Use hyphens to link the elements in compound adjectives that precede nouns:

a variable-angle accessory
increased cost-of-living raises

The hyphens in the second example prevent the reader from momentarily misinterpreting *increased* as an adjective modifying *cost* and *living* as a participle modifying *raises*.

A long string of compound adjectives can be confusing even if hyphens are used appropriately. To ensure clarity in such a case, put the adjectives into a clause or phrase following the noun:

UNCLEAR

an operator-initiated, default-prevention technique

CLEAR

a technique initiated by the operator for preventing default

In turning a string of adjectives into a phrase or clause, make sure the adjectives cannot be misread as verbs. Use the pronouns *that* and *which* to prevent confusion:

CONFUSING

The good experience provides is often hard to measure.

CLEAR

The good that experience provides is often hard to measure.

Maintain Number Agreement

Number disagreement commonly takes one of two forms in technical writing: (1) the verb disagrees in number with the subject when a prepositional phrase intervenes; (2) the pronoun disagrees in number with its antecedent or referent when the latter is a collective noun.

SUBJECT-VERB DISAGREEMENT

A prepositional phrase does not affect the number of the subject and the verb. The following examples show that the object of the preposition can be plural in a singular sentence or singular in a plural sentence. (The subjects and verbs are italicized.)

INCORRECT

The *result* of the tests *are* promising.

CORRECT

The *result* of the tests *is* promising.

INCORRECT

The *results* of the test *is* promising.

CORRECT

The *results* of the test *are* promising.

Don't be misled by the fact that the object of the preposition and the verb don't sound natural together, as in *tests is* or *test are*. Grammatical agreement of subject and verb is the primary consideration.

PRONOUN-REFERENT DISAGREEMENT

The problem of pronoun-referent disagreement crops up most often when the referent is a collective noun—one that can be interpreted as either singular or plural, depending on its usage:

INCORRECT

The *company* is proud to announce a new stock option plan for *their* employees.

CORRECT

The *company* is proud to announce a new stock option plan for *its* employees.

In this example, *the company* acts as a single unit; therefore, the singular verb, followed by a singular pronoun, is appropriate. When the individual members of a collective noun are stressed, however, plural pronouns and verbs are appropriate:

The inspection team have prepared their reports.

Punctuation

The Period

Periods are used in the following instances.

1. At the ends of sentences that do not ask questions or express strong emotion:

   The lateral stress still needs to be calculated.

2. After most abbreviations:

   M.D.
   U.S.A.
   etc.

(For a further discussion of abbreviations, see p. 189.)

The Exclamation Point

The exclamation point is used at the end of a sentence that expresses strong emotion, such as surprise or doubt:

> **The nuclear plant, which was originally expected to cost $1.6 billion, eventually cost more than $4 billion!**

Because technical writing requires objectivity and a calm, understated tone, technical writers rarely use exclamation points.

The Question Mark

The question mark is used at the end of a sentence that asks a direct question:

> **What did the commission say about effluents?**

Do not use a question mark at the end of a sentence that asks an indirect question:

> **He wanted to know whether the procedure had been approved for use.**

When a question mark is used within quotation marks, the quoted material needs no other end punctuation:

> **"What did the commission say about effluents?" she asked.**

The Comma

The comma is the most frequently used punctuation mark, as well as the one about whose usage many writers most often disagree. Following are the basic uses of the comma.

1. To separate the clauses of a compound sentence (one composed of two or more independent clauses) linked by a coordinating conjunction (*and, or, nor, but, so, for, yet*):

   > **Both methods are acceptable, *but* we have found that the Simpson procedure gives better results.**

   In many compound sentences, the comma is needed to prevent the reader from mistaking the subject of the second clause for an object of the verb in the first clause:

   > **The RESET command affects the field access, and the SEARCH command affects the filing arrangement.**

   Without the comma, the reader is likely to interpret the coordinating conjunction *and* as a simple conjunction linking *field access* and *SEARCH command.*

2. To separate items in a series composed of three or more elements:

   > **The manager of spare parts is responsible for ordering, stocking, and disbursing all spare parts for the entire plant.**

   The comma following the second-to-last item is required by most style manuals, despite the presence of the conjunction *and.* The comma clarifies the separation and prevents misunderstanding. For example, sometimes the second-to-last item will be a compound noun containing an *and.*

   > **The report will be distributed to Operations, Research *and* Development, *and* Accounting.**

3. To separate introductory words, phrases, and clauses from the main clause of the sentence:

   > **However, we will have to calculate the effect of the wind.**
   > **To facilitate trade, the government holds a yearly international conference.**
   > **Whether the workers like it or not, the managers have decided not to try the flextime plan.**

   In each of these three examples, the comma helps the reader follow the sentence. Note in the following example how the comma actually prevents misreading:

   > **Just as we finished eating, the rats discovered the treadmill.**

   The comma is optional if the introductory text is brief and cannot be misread.

CORRECT

First, let's take care of the introductions.

CORRECT

First let's take care of the introductions.

4. To separate the main clause from a dependent clause:

   The advertising campaign was canceled, although most of the executive council saw nothing wrong with it.
   Most accountants wear suits, whereas few engineers do.

5. To separate nonrestrictive modifiers (parenthetical clarifications) from the rest of the sentence:

   Jones, the temporary chairperson, called the meeting to order.

6. To separate interjections and transitional elements from the rest of the sentence:

   Yes, I admit your findings are correct.
   Their plans, however, have great potential.

7. To separate coordinate adjectives:

   The finished product was a sleek, comfortable cruiser.
   The heavy, awkward trains are still being used.

   The comma here takes the place of the conjunction *and*. If the adjectives are not coordinate—that is, if one of the adjectives modifies the combination of the adjective and the noun—do not use a comma:

   They decided to go to the first general meeting.

8. To signal that a word or phrase has been omitted from an elliptical expression:

   Smithers is in charge of the accounting; Harlen, the data management; Demarest, the publicity.

   In this example, the commas after "Harlen" and "Demarest" show that the phrase *is in charge of* has been omitted.

9. To separate a proper noun from the rest of the sentence in direct address:

   John, have you seen the purchase order from United?
   What I'd like to know, Betty, is why we didn't see this problem coming.

10. To introduce most quotations:

    He asked, "What time were they expected?"

11. To separate towns, states, and countries:

    Bethlehem, Pennsylvania, is the home of Lehigh University.
    Lee attended Lehigh University in Bethlehem, Pennsylvania, and the University of California at Berkeley.

    Note the use of the comma after "Pennsylvania."

12. To set off the year in dates:

    August 1, 1989, is the anticipated completion date.

    Note the use of the comma after 1989. If the month separates the date from the year, the commas are not used because the numbers are not next to each other:

    The anticipated completion date is 1 August 1989.

13. To clarify numbers:

    12,013,104

    (European practice reverses the use of commas and periods in writing numbers: periods are used to signify thousands, and commas to signify decimals.)

14. To separate names from professional or academic titles:

    Harold Clayton, Ph.D.
    Marion Fewick, CLU
    Joyce Carnone, P.E.

    Note that the comma also follows the title in a sentence:

    Harold Clayton, Ph.D., is the featured speaker.

COMMON ERRORS

1. No comma between the clauses of a compound sentence:

   INCORRECT

   The mixture was prepared from the two premixes and the remaining ingredients were then combined.

   CORRECT

   The mixture was prepared from the two premixes, and the remaining ingredients were then combined.

2. No comma (or just one comma) to set off a nonrestrictive modifier:

INCORRECT

The phone line, which was installed two weeks ago had to be disconnected.

CORRECT

The phone line, which was installed two weeks ago, had to be disconnected.

3. No comma separating introductory words, phrases, or clauses from the main clause, when misreading can occur:

INCORRECT

As President Canfield has been a great success.

CORRECT

As President, Canfield has been a great success.

4. No comma (or just one comma) to set off an interjection or a transitional element:

INCORRECT

Our new statistician, however used to work for Konaire, Inc.

CORRECT

Our new statistician, however, used to work for Konaire, Inc.

5. Comma splice (a comma used to "splice together" independent clauses not linked by a coordinating conjunction):

INCORRECT

All the motors were cleaned and dried after the water had entered, had they not been, additional damage would have occurred.

CORRECT

All the motors were cleaned and dried after the water had entered; had they not been, additional damage would have occurred.

CORRECT

All the motors were cleaned and dried after the water had entered. Had they not been, additional damage would have occurred.

6. Superfluous commas:

INCORRECT

Another of the many possibilities, is to use a "first in, first out" sequence. (*In this sentence, the comma separates the subject, "Another," from the verb, "is."*)

CORRECT

Another of the many possibilities is to use a "first in, first out" sequence.

INCORRECT

The schedules that have to be updated every month are, 14, 16, 21, 22, 27, and 31. (*In this sentence, the comma separates the verb from its complement.*)

CORRECT

The schedules that have to be updated every month are 14, 16, 21, 22, 27, and 31.

INCORRECT

The company has grown so big, that an informal evaluation procedure is no longer effective. (*In this sentence, the comma separates the predicate adjective "big" from the clause that modifies it.*)

CORRECT

The company has grown so big that an informal evaluation procedure is no longer effective.

INCORRECT

Recent studies, and reports by other firms confirm our experience. (*In this sentence, the comma separates the two elements in the compound subject.*)

CORRECT

Recent studies and reports by other firms confirm our experience.

INCORRECT

New and old employees who use the processed order form, do not completely understand the basis of the system. (*In this sentence, a comma separates the subject and its restrictive modifier from the verb.*)

CORRECT

New and old employees who use the processed order form do not completely understand the basis of the system.

The Semicolon

Semicolons are used in the following instances.

1. To separate independent clauses not linked by a coordinating conjunction:

   The second edition of the handbook is more up-to-date; however, it is more expensive.

2. To separate items in a series that already contains commas:

   The members elected three officers: Jack Resnick, president; Carol Wayshum, vice-president; Ahmed Jamoogian, recording secretary.

In this example, the semicolon acts as a "supercomma," keeping the names and titles clear.

COMMON ERROR

Use of a semicolon when a colon is called for:

INCORRECT

We still need one ingredient; luck.

CORRECT

We still need one ingredient: luck.

The Colon

Colons are used in the following instances.

1. To introduce a word, phrase, or clause that amplifies or explains a general statement:

   The project team lacked one crucial member: a project leader.
   Here is the client's request: we are to provide the preliminary proposal by November 13.
   We found three substances in excessive quantities: potassium, cyanide, and asbestos.

The week had been productive: fourteen projects had been completed and another dozen had been initiated.

Note that the text preceding a colon should be able to stand on its own as a main clause:

INCORRECT

We found: potassium, cyanide, and asbestos.

CORRECT

We found potassium, cyanide, and asbestos.

2. To introduce items in a vertical list, if the introductory text would be incomplete without the list:

   We found the following:
   potassium
   cyanide
   asbestos

3. To introduce long or formal quotations:

   The president began: "In the last year . . ."

COMMON ERROR

Use of a colon to separate a verb from its complement:

INCORRECT

The tools we need are: a plane, a level, and a T-square.

CORRECT

The tools we need are a plane, a level, and a T-square.

CORRECT

We need three tools: a plane, a level, and a T-square.

The Dash

Dashes are used in the following instances.

1. To set off a sudden change in thought or tone:

   The committee found—can you believe this?—that the company bore *full* responsibility for the accident.
   That's what she said—if I remember correctly.

2. To emphasize a parenthetical element:

   The managers' reports—all ten of them—recommend production cutbacks for the coming year.
   Arlene Kregman—the first woman elected to the board of directors—is the next scheduled speaker.

3. To set off an introductory series from its explanation:

   Wetsuits, weight belts, tanks—everything will have to be shipped in.

   When a series *follows* the general statement, a colon replaces the dash:

   Everything will have to be shipped in: wetsuits, weight belts, and tanks.

Note that typewriters and many word processors do not have a key for the dash. In typewritten or word-processed text, a dash is represented by two uninterrupted hyphens. No space precedes or follows the dash.

COMMON ERROR

Use of a dash as a "lazy" substitute for other punctuation marks:

INCORRECT

The regulations—which were issued yesterday—had been anticipated for months.

CORRECT

The regulations, which were issued yesterday, had been anticipated for months.

INCORRECT

Many candidates applied—however, only one was chosen.

CORRECT

Many candidates applied; however, only one was chosen.

Parentheses

Parentheses are used in the following instances.

1. To set off incidental information:

   Please call me (X3104) when you get the information.

   Galileo (1564–1642) is often considered the father of modern astronomy.
   H. W. Fowler's *Modern English Usage* (New York: Oxford University Press, 2d ed., 1965) is still the final arbiter.

2. To enclose numbers and letters that label items listed in a sentence:

   To transfer a call within the office, (1) place the party on HOLD, (2) press TRANSFER, (3) press the extension number, and (4) hang up.

Use both a left and a right parenthesis—not just a right parenthesis—in this situation.

COMMON ERROR

Use of parentheses instead of brackets to enclose the writer's interruption of a quotation (see the discussion of brackets):

INCORRECT

He said, "The new manager (Farnham) is due in next week."

CORRECT

He said, "The new manager [Farnham] is due in next week."

The Hyphen

Hyphens are used in the following instances.

1. In general, to form compound adjectives that precede nouns:

   general-purpose register
   meat-eating dinosaur
   chain-driven saw

   Note that hyphens are not used after adverbs that end in *-ly*:

   newly acquired terminal

   Also note that hyphens are not always used when the compound adjective follows the noun:

   The Woodchuck saw is chain driven.

   Many organizations have their own preferences about hyphenating compound adjectives. Check to see if your organization has a preference.

2. To form some compound nouns:

   editor-in-chief
   member-at-large

3. To form fractions and compound numbers:

   one-half
   fifty-six

4. To attach some prefixes and suffixes:

   post-1945
   president-elect

5. To divide a word at the end of a line:

   We will meet in the pavil-
   ion in one hour.

   Whenever possible, avoid such breaks; they annoy some readers. When you do use them, check the dictionary to make sure you have divided the word *between* syllables.

The Apostrophe

Apostrophes are used in the following instances.

1. To indicate the possessive case:

   the manager's goals
   the supervisor's lounge
   the employees' credit union
   Charles's T-square

   For joint possession, add the apostrophe and the *s* to only the last noun or proper noun:

   Watson and Crick's discovery

   For separate possession, add an apostrophe and an *s* to each of the nouns or pronouns:

   Newton's and Galileo's ideas

   Make sure you do not add an apostrophe or an *s* to possessive pronouns: *his, hers, its, ours, yours, theirs.*

2. To form contractions:

   I've
   can't
   shouldn't
   it's

   The apostrophe usually indicates an omitted letter or letters. For example, *can't* is *can(no)t, it's* is *it(i)s.*
   Some organizations discourage the use of contractions; others have no preference. Find out the policy your organization follows.

3. To indicate special plurals:

   three 9's
   two different JCL's
   the why's and how's of the problem

   As in the case of contractions, it is a good idea to learn the stylistic preferences of your organization. Usage varies considerably.

COMMON ERROR

Use of the contraction *it's* in place of the possessive pronoun *its*.

INCORRECT

The company does not feel that the problem is *it's* responsibility.

CORRECT

The company does not feel that the problem is *its* responsibility.

Quotation Marks

Quotation marks are used in the following instances.

1. To indicate titles of short works, such as articles, essays, or chapters:

   Smith's essay "Solar Heating Alternatives"

2. To call attention to a word or phrase that is being used in an unusual way or in an unusual context:

   A proposal is "wired" if the sponsoring agency has already decided who will be granted the contract.

   Don't use quotation marks as a means of excusing poor word choice:

   The new director has been a real "pain."

3. To indicate direct quotation: that is, the words a person has said or written:

   "In the future," he said, "check with me before authorizing any large purchases."
   As Breyer wrote, "Morale *is* productivity."

   Do not use quotation marks to indicate indirect quotation:

   INCORRECT

   He said that "third-quarter profits would be up."

   CORRECT

   He said that third-quarter profits would be up.

   CORRECT

   He said, "Third-quarter profits will be up."

RELATED PUNCTUATION

Note that if the sentence contains a "tag"—a phrase identifying the speaker or writer—a comma is used to separate it from the quotation:

John replied, "I'll try to fly out there tomorrow."
"I'll try to fly out there tomorrow," John replied.

Informal and brief quotations require no punctuation before the quotation marks:

She said "Why?"

In the United States (but not in most other English-speaking nations), commas and periods at the ends of quotations are placed within the quotation marks:

The project engineer reported, "A new factor has been added."
"A new factor has been added," the project engineer reported.

Question marks, dashes, and exclamation points, on the other hand, are placed inside the quotation marks when they apply only to the quotation and outside the quotation marks when they apply to the whole sentence:

He asked, "Did the shipment come in yet?"
Did he say, "This is the limit"?

Note that only one punctuation mark is used at the end of a set of quotation marks:

INCORRECT

Did she say, "What time is it?"?

CORRECT

Did she say, "What time is it?"

Block Quotations

When quotations reach a certain length—generally, more than four lines—writers tend to switch to a block format. In typewritten manuscript, a block quotation is usually

1. indented ten spaces from the left-hand margin
2. single-spaced
3. typed without quotation marks
4. introduced by a colon

(Different organizations observe their own variations on these basic rules.)

McFarland writes:

> **The extent to which organisms adapt to their environment is still being charted. Many animals, we have recently learned, respond to a dry winter with an automatic birth-control chemical that limits the number of young to be born that spring. This prevents mass starvation among the species in that locale.**

Hurlihy concurs. She writes, "Biological adaptation will be a major research area during the next decade."

Mechanics

Ellipses

Ellipses (three spaced periods) indicate the omission of some material from a quotation. A fourth period with no space before it precedes ellipses when the sentence in the source has ended and you are omitting material that follows, or when the omission follows a portion of the source's sentence that is in itself a grammatically complete sentence:

"Send the updated report . . . as soon as you can."
Larkin refers to the project as "an attempt . . . to clarify the issue of compulsory arbitration. . . . We do not foresee an end to the legal wrangling . . . but perhaps the report can serve as a definition of the areas of contention."

In the second example, the writer has omitted words after *attempt* and after *wrangling.* In addition, after *arbitration,* which ends the original writer's sentence, she has omitted a full sentence.

Brackets

Brackets are used in the following instances.

1. To indicate words added to a quotation:

   The minutes of the meeting note that "He [Pearson] spoke out against the proposal."

   A better approach would be to alter the quotation:

   Pearson "spoke out against the proposal."

2. To indicate parentheses within parentheses:

   (For further information, see Charles Houghton's *Civil Engineering Today* [New York: Arch Press, 1987].)

Italics

Italics (or underlining) are used in the following instances.

1. For words used as words:

   In this report, the word *operator* will refer to any individual who is actually in charge of the equipment, regardless of that individual's certification.

2. To indicate titles of long works (books, manuals, etc.), periodicals and newspapers, films, plays, and long musical works:

   See Houghton's *Civil Engineering Today.*
   We subscribe to the *New York Times.*

3. To indicate the names of ships, trains, and airplanes:

   The shipment is expected to arrive next week on the *Penguin.*

4. To set off foreign expressions that have not become fully assimilated into English:

   The speaker was guilty of *ad hominem* arguments.

5. To emphasize words or phrases:

   ***Do not* press the ERASE key.**

   If your typewriter or word processor does not have italic type, indicate italics by underlining.

   Darwin's Origin of Species is still read today.

Numbers

The use of numbers varies considerably. Therefore, you should find out what guidelines your organization or research area follows in choosing between words and numerals. Many organizations use the following guidelines.

1. Use numerals for technical quantities, especially if a unit of measurement is included:

   3 feet
   12 grams
   43,219 square miles
   36 hectares

2. Use numerals for nontechnical quantities of 10 or more:

   300 persons
   12 whales
   35% increase

3. Use words for nontechnical quantities of less than 10:

   three persons
   six whales

4. Use both words and numerals

   a. For back-to-back numbers:

      six 3-inch screws
      fourteen 12-foot ladders
      3,012 five-piece starter units

In general, use the numeral for the technical unit. If the nontechnical quantity would be cumbersome in words, use the numeral.

b. For round numbers over 999,999:

14 million light years
$64 billion

c. For numbers in legal contracts or in documents intended for international readers:

thirty-seven thousand dollars ($37,000)
five (5) relays

d. For addresses:

3801 Fifteenth Street

SPECIAL CASES

1. If a number begins a sentence, use words, not numerals:

   Thirty-seven acres was the agreed-upon size of the lot.

   Many writers would revise the sentence to avoid this problem:

   The agreed-upon size of the lot was 37 acres.

2. Don't use both numerals and words in the same sentence to refer to the same unit:

   On Tuesday the attendance was 13; on Wednesday, 8.

3. Write out fractions, except if they are linked to technical units:

   two-thirds of the members
   3½ hp

4. Write out approximations:

   approximately ten thousand people
   about two million trees

5. Use numerals for figures and tables and for page numbers:

   Figure 1
   Table 13
   page 261

6. Use numerals for decimals:

   3.14
   1,013.065

   Add a zero before decimals of less than one:

   0.146
   0.006

7. Avoid expressing months as numbers, as in "3/7/88": in the United States, this means March 7, 1988; in many other countries, it means July 3, 1988. Use one of the following forms:

   March 7, 1987
   7 March 1987

8. Use numerals for times if A.M. or P.M. is used:

   6:00 A.M.
   six o'clock

Abbreviations

Abbreviations provide a useful way to save time and space, but you must use them carefully; you can never be sure that your readers will understand them. Many companies and professional organizations have lists of approved abbreviations.

Analyze your audience in determining whether and how to abbreviate. If your readers include nontechnical people unfamiliar with your field, either write out the technical terms or attach a list of abbreviations. If you are new in an organization or are writing for publication for the first time in a certain field, find out what abbreviations are commonly used. If for any reason you are unsure about whether or how to abbreviate, write out the word.

The following are general guidelines about abbreviations.

1. You may make up your own abbreviations. For the first reference to the term, write it out and include, parenthetically, the abbreviation. In subsequent references, use the abbreviation. For long works, you might want to write out the term at the start of major units, such as chapters.

   The heart of the new system is the self-loading cartridge (slc).

   This technique is also useful, of course, in referring to existing abbreviations that your readers might not know:

   The cathode-ray tube (CRT) is your control center.

2. Most abbreviations do not take plurals:

   1 lb
   3 lb

3. Most abbreviations in scientific writing are not followed by periods:

   lb
   cos
   dc

   If the abbreviation can be confused with another word, however, use a period:

   in.
   Fig.

4. Spell out the unit if the number preceding it is spelled out or if no number precedes it:

   How many square meters is the site?

Capitalization

For the most part, the conventions of capitalization in general writing apply in technical writing.

1. Capitalize proper nouns, titles, trade names, places, languages, religions, and organizations:

   William Rusham
   Director of Personnel
   Quick Fix Erasers
   Bethesda, Maryland
   Methodism
   Italian
   Society for Technical Communication

   In some organizations, job titles are not capitalized unless they refer to specific persons:

   Alfred Loggins, Director of Personnel, is interested in being considered for vice-president of marketing.

2. Capitalize headings and labels:

   A Proposal to Implement the Wilkins Conversion System
   Section One
   The Problem
   Figure 6
   Mitosis
   Table 3
   Rate of Inflation, 1980–1985

Appendix B
How to Document Sources

Author/Date Citations and Bibliography

Numbered Citations and Numbered Bibliography

Documentation is the explicit identification of the sources of the ideas and quotations used in your document.

For you as the writer, complete and accurate documentation is primarily a professional obligation—a matter of ethics. But it also serves to substantiate the document. Effective documentation helps to place the writing within the general context of continuing research and to define it as a responsible contribution to knowledge in the field. Failure to document a source—whether intentionally or unintentionally—is plagiarism. At most universities and colleges, plagiarism means automatic failure of the course and, in some instances, suspension or dismissal. In many companies, it is grounds for immediate dismissal.

For your readers, complete and accurate documentation is an invaluable tool. It enables them to find the source you have relied on, should they want to read more about a particular subject.

What kind of material should be documented? Any quotation from a printed source or an interview, even if it is only a few words, should be documented. In addition, a paraphrased idea, concept, or opinion gathered from your reading should be documented. There is one exception to this rule: if the idea or concept is so well known that it has become, in effect, general knowledge, it need not be documented. Examples of knowledge that is within the public domain would include Einstein's theory of relativity and the Laffer curve. If you are unsure whether an item is within the public domain, document it anyway, just to be safe.

Many organizations have their own preferences for documentation style; others use published style guides, such as the *U.S. Government Printing Office Style Manual*, the American Chemical Society's *Handbook for Authors*, or *The Chicago Manual of Style*. You should find out what your organization's style is and abide by it.

The system of documentation you are probably most familiar with—the notes and bibliography style—is used rarely today, especially outside the liberal arts. Even the Modern Language Association, a major organization in the humanities, no longer recommends the use of the notes-and-bibliography style because of the duplication involved in creating two different lists.

Two basic systems of documentation have become standard:

1. author/date citations and bibliography
2. numbered citations and numbered bibliography

Variations on these systems abound.

Author/Date Citations and Bibliography

In the author/date style, a parenthetical notation that includes the name of the source's author and the date of its publication immediately follows the quoted or paraphrased material:

```
This phenomenon was identified as early as forty
years ago (Wilkinson 1943).
```

Sometimes, particularly if the reference is to a specific fact or idea, the page (or pages) from the source is also listed:

```
(Wilkinson 1943: 36–37)
```

A citation can also be integrated into the sentence:

```
Wilkinson (1943: 36–37) identified this phenomenon
as early as forty years ago.
```

If two or more sources by the same author in the same year are listed in the bibliography, the notation may include an abbreviated title, to prevent confusion:

```
(Wilkinson, "Cornea Research," 1943: 36–37)
```

Or the citation for the first source written that year can be identified with a lowercase letter:

```
(Wilkinson 1943a: 36–37)
```

The second source would be identified similarly:

```
(Wilkinson 1943b: 19–21)
```

The simplicity and flexibility of the author/date system make it highly attractive. Of course, because the citations are minimal and because their form is dictated more by common sense than by a style sheet, a conventional bibliography that contains complete publication information must be used in conjunction with them, following the text. Your obligation as a writer when you are using this system is to leave your reader no doubt about which of your many sources you are citing in any particular instance. *The Chicago Manual of Style*, 13th edition (Chicago: University of Chicago Press, 1982), advocates using the author/date system throughout the natural and social sciences and recommends the following bibliographic conventions for use with author/date citations.

The order of basic information included in this style of bibliography is as follows:

FOR BOOKS

1. author (last name followed by initials)
2. date of publication
3. title
4. place of publication
5. publisher

FOR ARTICLES

1. author
2. date of publication
3. article title
4. journal or anthology title
5. volume (of a journal) or place of publication and publisher (of an anthology)
6. inclusive pages

The individual entries are arranged alphabetically by author (and then by date if two or more works by the same author are listed) within the bibliography. Anonymous works are integrated into the alphabetical listing by title. Where several works by the same author are included, they are arranged by date under the author's name. In such cases, a long dash (10 hyphens on a typewriter) takes the place of the author's name in all entries after the first. The first line of a bibliographic entry is flush left with the margin; each succeeding line is indented (five spaces on a typewriter).

Chapman, D. L. 1988. The closed frontier: Why Detroit can't make cars that people will buy. *Motorist's Metronome* 12(June):17–26.

----------. 1985. *The driver's guide to evaluating compact automobiles.* Athens, Ga.:Consumer Press.

Courting credibility: Detroit and its mpg figures. 1986. *Countercultural Car & Driver* (July):19.

Following are the standard bibliographic forms used with various types of sources.

A BOOK

Cunningham, W. 1980. *Crisis at Three Mile Island.* New York: Madison.

The author's surname is followed by the first initial only. The date of publication comes next, followed by the title of the book, underlined (italicized in print). In the natural and social sciences, the style is generally to capitalize only the first letter of the title, the subtitle (if there is one), and all proper nouns. The last items are the location and name of the publisher.

For a book by two or more authors, all of the authors are named.

Cunningham, W., and A. Breyer . . .

Only the name of the first author is inverted.

A BOOK ISSUED BY AN ORGANIZATION

Department of Energy. 1979. *The energy situation in the eighties*. Washington, D.C.: U.S. Government Printing Office, Technical Report 11346–53.

A BOOK COMPILED BY AN EDITOR OR ISSUED UNDER AN EDITOR'S NAME

Morgan, K. E., ed. 1987. *Readings in alternative energies*. Boston: Smith-Howell.

AN EDITION OTHER THAN THE FIRST

Schonberg, N. 1987. *Solid state physics*. 3d ed. London: Paragon.

AN ARTICLE INCLUDED IN A BOOK

May, B., and J. Deacon. 1986. Amplification

systems. In Third Annual Conference of the American Electronics Association, ed. A. Kooper, 101–14. Miami: Valley Press.

A JOURNAL ARTICLE

Hastings, W. 1983. The Skylab debate. The Modern Inquirer 13:311–18.

The title of the journal is underlined (italicized in print). The first letters of the first and last words and all nouns, adjectives, verbs (except the infinitive *to*), adverbs, and subordinate conjunctions are capitalized. After the journal title comes the volume and page numbers of the article.

AN ANONYMOUS JOURNAL ARTICLE

The state of the art in microcomputers. 1985. Newscene 56:406–21.

Anonymous journal articles are arranged alphabetically by title. If the title begins with a grammatical article such as *the* or *a*, alphabetize it under the first word following the article (in this case, *state*).

A NEWSPAPER ARTICLE

Eberstadt, A. 1987. Why not a Rabbit, why not a Fox? Morristown Mirror and Telegraph. Sunday 31 July: Business and Finance Section, p. 6.

A PERSONAL INTERVIEW

Riccio, Dr. Louis, Professor of Operations Research, Tulane University. Interview with author. New York City, 13 July 1986.

The name and professional title of the respondent should be cited first, then the fact that the source was an interview, and finally the place and date of the interview.

A QUESTIONNAIRE CONDUCTED BY SOMEONE OTHER THAN THE AUTHOR

Recycled Resources Corp. Data derived from questionnaire administered to 33 supervisors in the Bethlehem plant, November 3–4, 1987.

An individual, an outside firm, or a department might also serve as the questionnaire's "author" for bibliographic purposes. If you are citing data derived from your own questionnaire, a bibliographic entry is unnecessary: you include the questionnaire and appropriate background information in an appendix. When citing your own questionnaire data in the text of your report, refer your readers to the appendix.

COMPUTER SOFTWARE

Block, K. 1987. Planner. Computer software. Global Software. IBM-PC.

Tools for Drafting. 1986. Computer software. Software International. CP/M 2.3.

Begin the citation with the author of the software if an author is identified. Underline (or italicize) the name of the program, and then identify the program as computer software. List the name of the publisher. Finally, add any identifying information, such as the kind of system or the brand and model of hardware the software runs on.

INFORMATION RETRIEVED FROM A DATA-BASE SERVICE

Crayton, H. 1987. "A new method for insulin implants." Biotechnology 6: 31–42. DIALOG file 17, item 230043 867745.

Information retrieved from a data-base service is treated like a printed source. At the end of the citation, however, you should list the name of the data-base service and the necessary identifying information to enable a reader to find the item.

Following is a sample bibliography for use with the author/date citation system:

BIBLIOGRAPHY

Daly, P. H. 1984. Selecting and designing a group insurance plan. Personnel Journal 54:322–23.

Flanders, A. 1984. Measured daywork and collective bargaining. British Journal of Industrial Relations 9:368–92.

Goodman, R. K., J. H. Wakely, and R. H. Ruh. 1983. What employees think of the Scanlon Plan. Personnel 6:22–29.

Trencher, P. 1984. Recent trends in labor-management relations. New York Madison.

Zwicker, D. Professor of Industrial Relations, Hewlett College. Interview with author. Philadelphia, 19 March 1985.

Numbered Citations and Numbered Bibliography

You will occasionally encounter a variation on the author/date notation system—the numbered citation and numbered bibliography system. In this system, the items in the bibliography are arranged either alphabetically or in order of the first appearance of each source in the text and then assigned a sequential number. The citation is the number of the source, enclosed in brackets or parentheses:

According to Hodge [3], "There is always the danger that the soil will shift." However, no shifts have been noted.

As in author/date citations, page numbers may be added:

According to Hodge [3:26], "There is always . . ."

In this example, *3* means that Hodge is the third item in the numbered bibliography; *26* is the page number.

Except that they are numbered, the individual bibliography entries in this system are identical to those used with the author/date system.

Appendix C
Word Processing and the Writing Process

Preparing, Research, and Organizing

Drafting

Revising

Most people who have never used a word processor are skeptical when they hear stories about how terrific it is. How can a machine increase a writer's productivity by 100 percent, 200 percent, or more? How can a machine eliminate misspellings and grammatical mistakes? How can a machine make it fun to write?

These claims, and dozens more, are made every day. Some of them are merely advertising talk, but others are true. You can write faster and more easily with a word processor. You can catch more errors. You can produce neater and more professional-looking documents. There is only one problem: the word processor won't tell you what to say. You still have to do the writing.

In the following discussion of word processors and the writing process, the term *word processor* will be used to refer to dedicated word processors (computers especially designed for word processing), microcomputers used with word-processing programs, and larger computers accessed through remote terminals.

Word processors offer many advantages over pen and paper at every stage of the writing process, whether it be preparing, research, organizing, drafting, or revising.

Preparing, Research, and Organizing

A number of idea-generating programs exists that appear to help many students create topics to write about. These programs ask you to respond to a series of questions about a subject area of your choice. The questions force you to think about the audience and the purpose of the report and to indicate the kinds of information you will need to write it. There is no scientific proof that using these programs is more effective than simply answering the same questions on paper. However, some scholars report that as their students use these idea-generating programs several times, they learn how to ask the right questions and therefore write better reports.

Once you sit down to brainstorm, the word processor will help you save time. Because even slow typists can work quickly on a word processor, you can create a full brainstorming list quickly and easily—and everything you write will be easy to read.

A second use of word processors in the research stage is in taking notes. If you can bring your source material to the word processor, you can save time and trouble later. Rather than take notes on index cards, you type them into the word processor. When you want to use that information in your document, you can put it there easily and revise it any way you want. Making your bibliography will be much easier, too, and you won't introduce new errors, because you'll simply move the entries into position without having to type them in from handwritten notes.

Moving items from one position to another on the screen is effortless; therefore, you are more likely to try out different alternatives as you classify items into groups. The same holds true when you sequence the groups. Of course you can cut and paste on paper, but the process is much more cumbersome.

Drafting

The word processor is a useful tool during the drafting stage because it encourages you to write faster. Knowing that you can easily move information from one place in the text to another gives you the freedom to begin writing in the middle, on a technical point you are familiar with. When you find that you can't think of what to write on one subject, just skip a few lines and go to the next subject on your outline. Shuffling your text later takes only a few seconds.

With a word processor, you can easily write your draft right on your outline. The advantage is that you are less likely to lose sight of your overall plan. Before you start to draft, make a copy of your outline. You'll be able to use it later to create a table of contents. On the other copy of the outline, start to draft. As you write, the portion of the outline below will scroll downwards to accommodate your draft. You don't have a separate outline and a separate draft: the outline becomes the draft.

Because word processors are relatively quiet and easy to type on, you will probably find that you can generate much more writing in a given period of time. You don't have the physical effort involved in writing by hand, and you don't have to return the carriage at the end of the line as you do on a typewriter. Producing a lot of writing quickly is exactly the point of drafting: you want to have material to revise later.

Another factor that encourages speed and productivity is that you don't worry about the quality of the writing or about typographical errors. You can concentrate on what you are trying to say because making changes, whether large or small, requires so little effort.

Some writers find that the word processor provides an effective way to break the time-consuming habit of revising what they have just typed. The screen has a contrast knob, just as a television set does. Turning the contrast

knob all the way makes the screen black. This technique, called invisible writing, encourages writers to close their eyes or look at the keyboard. The result is that they don't stop typing so often.

A common feature on word processors—the search-and-replace function—also increases the writer's speed during the drafting stage. The search-and-replace function lets you find any phrase, word, or characters and replace them with any other writing. For example, if you need to use the word *potentiometer* a number of times in your document, you can simply type in *po* each time. Then, during the revising stage, you can instruct the word processor to change every *po* to *potentiometer*. This use of the search-and-replace function not only saves time as you draft; it also reduces the chances of misspelling, for you have to spell the word correctly only once.

Revising

Word processors make every kind of revising easier. Most obviously, your writing is legible, so you see what you've done without being distracted by sloppy handwriting. And because your writing is typed neatly, you have a more objective perspective on your work. You are looking at it as others will.

If the obvious typographical errors distract you, fix them first so that you can concentrate on substantive changes. All word processors have easy-to-use add and delete functions so that you change *hte* to *the* in a second.

You can easily make major revisions to the structure and organization of the document. All word processors let you move text—anything from a single letter to whole paragraphs—simply and quickly. Therefore, you can try out different versions of the document without the bother of cutting and pasting pieces of paper. Most word processors also have a copy function, which lets you copy text—such as an introductory paragraph—and move the copy to some other location without moving the original text. With the copy function, you can in effect create two different versions of the document simultaneously and decide which one works better.

A number of different editing programs are available that help you identify problem areas that need to be fixed.

One common editing program is called a spelling checker, which compares what you have typed with a dictionary, usually of 20,000 to 90,000 words. The program can check approximately five thousand to ten thousand words per minute and will alert you when it sees a word that isn't in its dictionary. Although that word might be misspelled, it might be a correctly spelled word that isn't in the dictionary. You can add the word to your dictionary so that the program will know in the future that the word is not misspelled. If the word is misspelled, you have to look it up in a dictionary. Without the spelling checker to point out the error, you might not have known the word was misspelled.

One limitation of any spelling-checker program is that it cannot tell whether you have used the correct word; it can tell you only whether the word you have used is in its dictionary. Therefore, if you have typed, "We need too dozen test tubes," the spelling checker will not see a problem.

A related program is a thesaurus, which lists similar words for many common words. A thesaurus program has the same strengths and weaknesses as a printed thesaurus: if you know the word you are looking for but can't quite think of it, the thesaurus will help you remember it. But the terms listed might not be closely enough related to the key term to function as synonyms. Unless you are aware of the shades of difference, you might be tempted to substitute an inappropriate word. For example, the word *journal* is followed in *Roget's College Thesaurus* by the word *diary*. A personal journal is a diary, but a professional periodical certainly isn't.

A word-usage program measures the frequency of particular words and the length of words. All of us overuse some words, but without the word processor we have a difficult time determining which ones. Word length is a useful factor to know, for technical terms frequently are long words. After we analyze our audience and purpose, we need to consider the amount of technical terminology to include in the document.

And finally, there are style programs, many of which perform several functions. For instance, they count such factors as sentence length, number of passive-voice constructions and grammatical expletives, and types of sentence (see Chapter 3). Many style programs identify abstract words and suggest more specific ones. Several point out sexist terms and provide nonsexist alternatives. Many point out fancy words, such as *demonstrate*, and suggest substitutes, such as *show*. And finally, a number apply readability formulas to the text—indexes of how difficult it is to read.

Keep in mind just what these different programs can do and what they cannot do. They can point out your use of the passive voice, but they cannot tell you whether the passive voice is preferable to the active voice in a particular sentence. These different style programs in a way make your job as a writer more challenging; by pointing out potential problems, they force you to make decisions about issues that you might not have even noticed

without the word processor. But the payoff is that a wise use of the programs will give you a better document.

Many of the functions carried out by these style programs can be done instead with the search function. For instance, if you know that you overuse expletives ("It is . . . ," "there is . . . ," "there are . . . "), you can search for words such as *is* and *there*. Some of the uses of these words, of course, will not be expletives, but you can revise those that are inappropriate. Or if you realize you overuse nominalizations (noun forms of verbs, such as *installation* for *install*), you can search for the common suffixes (such as *-tion*, *-ance*, and *-ment*) used in nominalizations.

Although word processors can help you do much of the work involved in revision, they cannot replace a careful reading by another person. Revision programs will calculate sentence length more accurately than a person can, but they cannot identify unclear explanations, contradictions, inaccurate data, inappropriate choice of vocabulary, and so forth. Use the revision programs, revise your document yourself, and then get help from someone you trust.

Selected Bibliography

Technical Writing

Currently, more than one hundred technical writing texts and guides are on the market. Following is a representative selection.

Blicq, R. S. 1986. *Technically—write! Communicating in a technological era.* 3d. ed. Englewood Cliffs, N.J.: Prentice-Hall.

Brusaw, C. T., G. J. Alred, and W. E. Oliu. 1987. *Handbook of technical writing.* 3d ed. New York: St. Martin's. used in industry

Houp, K. W., and T. E. Pearsall. 1984. *Reporting technical information.* 5th ed. New York: Macmillan.

Kolin, P. C., and J. Kolin. 1985. *Models for technical writing.* New York: St. Martin's.

Lannon, J. M. 1984. *Technical writing.* 3d ed. Boston: Little, Brown.

Markel, M. H. 1987. *Technical writing: situations and strategies.* 2d ed. 1988. New York: St. Martin's.

Mathes, J. C., and D. W. Stevenson. 1976. *Designing technical reports.* Indianapolis: Bobbs-Merrill.

Mills, G. H., and J. A. Walter. 1978. *Technical writing.* 4th ed. New York: Holt, Rinehart and Winston.

Pickett, N. A., and A. A. Laster. 1984. *Technical English.* 4th ed. New York: Harper & Row.

Sherman, T. A., and S. S. Johnson. 1983. *Modern technical writing.* 4th ed. Englewood Cliffs, N.J.: Prentice-Hall.

Wiseman, H. M. 1985. *Basic technical writing.* 5th ed. Columbus, Ohio: Charles E. Merrill.

Also see the following journals:

IEEE Transactions on Professional Communication
Journal of Business Communication
Journal of Technical Writing and Communication
Technical Communication
The Technical Writing Teacher

Usage and General Writing

Barzun, J. 1985. *Simple and direct: A rhetoric for writers.* New York: Harper & Row.

Bernstein, T. M. 1965. *The careful writer.* New York: Atheneum.

Corbett, E. P. J. 1971. *Classical rhetoric for the modern student.* 2d ed. New York: Oxford University Press.

Flesch, R. 1980. *The ABC of style—A guide to plain English.* New York: Harper & Row.

Fowler, H. W. 1965. *A dictionary of modern English usage.* 2d ed., rev. by Sir E. Gowers. New York: Oxford University Press.

Hayakawa, S. I. 1978. *Language in thought and action.* 4th ed. New York: Harcourt Brace Jovanovich.

Miller, K., and K. Swift. 1980.*The handbook of nonsexist writing.* New York: Lippincott & Crowell.

Sorrels, B. D. 1983. *The nonsexist communicator: Solving the problems of gender and awkwardness in modern English.* Englewood Cliffs, N.J.: Prentice-Hall.

Strunk, W., and E. B. White. 1979. *The elements of style.* 3d ed. New York: Macmillan.

Trimble, J. R. 1975. *Writing with style: Conversations on the art of writing.* Englewood Cliffs, N.J.: Prentice-Hall.

Style Manuals

American Chemical Society. 1978. *Handbook for authors.* Washington, D.C.: American Chemical Society.

American National Standards Institute. 1979. *American national standard for the preparation of scientific papers for written or oral preparation.* ANSI Z39.16—1972. New York: American National Standards Institute.

CBE Style Manual Committee. 1983. *Council of biology editors style manual: A guide for authors, editors, and publishers in the biological sciences.* 5th ed. Washington, D.C.: Council of Biology Editors.

The Chicago manual of style. 1982. 13th ed., rev. Chicago: University of Chicago Press.

ORNL style guide. 1974. Oak Ridge, Tenn.: Oak Ridge National Laboratory.

Pollack, G. 1977. *Handbook for ASM editors.* Washington, D.C.: American Society for Microbiology.

Publications manual of the American Psychological Association. 1983. 3d ed. Washington, D.C.: American Psychological Association.

Skillin, M., R. Gay, et al. 1974. *Words into type.* 3d ed. Englewood Cliffs, N.J.: Prentice-Hall.

U.S. Government Printing Office style manual. 1967. Rev. ed. Washington, D.C.: Government Printing Office.

Also, many private corporations, such as John Deere, DuPont, Ford Motor Company, General Electric, and Westinghouse, have their own style manuals.

Graphic Aids

Beakley, G. C., Jr., and D. D. Autore. 1973. *Graphics for design and visualization.* New York: Macmillan.

Lefferts, R. 1982. *How to prepare charts & graphs for effective reports.* New York: Harper & Row.

MacGregor, A. J. 1979. *Graphics simplified: How to plan and prepare effective charts, graphs, illustrations, and other visual aids.* Toronto: University of Toronto Press.

Morris, G. E. 1975. *Technical illustrating.* Englewood Cliffs, N.J.: Prentice-Hall.

Thomas, T. A. 1978. *Technical illustration*. 3d ed. New York: McGraw-Hill.
Tufte, E. R. 1983. *The visual display of quantitative information*. Cheshire, Conn.: Graphics Press.
Turnbull, A. T., and R. N. Baird. 1980. *The graphics of communication*. 4th ed. New York: Holt, Rinehart and Winston.

Also see the following journals:
Graphic Arts Monthly
Graphics: USA

Proposals

Lefferts, R. 1983. *The basic handbook of grants management*. New York: Basic Books.
Society for Technical Communication. 1973. *Proposals and their preparation*. Vol. 1. Washington, D.C.: Society for Technical Communication.

Oral Presentations

Anastasi, T. E., Jr. 1972. *Communicating for results*. Menlo Park, Calif.: Cummings.
Howell, W. S., and E. G. Barmann. 1971. *Presentational speaking for business and the professions*. New York: Harper & Row.
Weiss, H., and J. B. McGrath, Jr. 1963. *Technically speaking: Oral communication for engineers, scientists, and technical personnel*. New York: McGraw-Hill.

Word Processing

Bernstein, S., and L. McGarry. 1986. *Making art on your computer*. New York: Watson-Guptill.
Fluegelman, A., and J. J. Hewes. 1983. *Writing in the computer age: Word processing skills and style for every writer*. Garden City, N.Y.: Anchor Press/Doubleday.
Hyman, M. I. 1985. *Advanced IBM PC graphics: state of the art*. New York: Brady Communications.
McCunn, D. 1982. *Write, edit, & print: Word processing with personal computers*. San Francisco: Design Enterprises of San Francisco.
Price, J., and Pinneau Urban, L. 1984. *The definitive word-processing book*. New York: Viking Penguin.
Schwartz, H. J. 1985. *Interactive writing: Composing with a word processor*. New York: Holt, Rinehart and Winston.
Stern, F. 1983. *Word processing and beyond*. Santa Fe, N.M.: John Muir.
Stone, M. D. 1984. *Getting on-line: A guide to accessing computer information services*. Englewood Cliffs, N.J.: Prentice-Hall.
Zinsser, W. 1983. *Writing with a word processor*. New York: Harper & Row.

Index